Marcus Chown is an award-winning writer and broadcaster. Formerly a radio astronomer at the California Institute of Technology in Pasadena, he is now cosmology consultant of the UK weekly science magazine *New Scientist*. His books include *What a Wonderful World*, *Quantum Theory Cannot Hurt You* and *We Need to Talk about Kelvin* (shortlisted for the 2010 Royal Society Science Book Prize). Marcus is also author of *Solar System for iPad*, which won *The Bookseller* 2011 Digital Innovation of the Year award.

www.marcuschown.com

@marcuschown

Praise for *The Ascent of Gravity*

'Timely, accessible and peppered with quotes from Douglas Adams and Terry Pratchett, this history of something we all feel but still cannot quite grasp has an admirably light touch'
The Sunday Times, Science Book of the Year

'[An] entertaining and at times mind-boggling guide to the weakest of nature's fundamental forces, which also controls the fate of the universe' Manjit Kumar, *The Times*

'[Chown] tells his story clearly and sets out the key ideas without recourse to jargon and intimidating mathematics . . . Eminently readable' Graham Farmelo, *Guardian*

'Genial wit and scientific flair awaits' *Nature*

'Marcus Chown is one of the UK's best writers on physics and astronomy – it's excellent to see him back on what he does best . . . no one has covered the topic with such a light touch and joie de vivre as Chown . . . A very readable exploration of humanity's gradual realisation of what gravity was about with all of Chown's usual sparkle . . . a delight'
Brian Clegg, popularscience.co.uk

'Compact and accessible while remaining comprehensive. A welcome addition to anyone's popular science library, written in a relaxed style and full of relevant quotations'

BBC Sky at Night Magazine

'A readable romp through the history of cosmology and its possible future, all tied together through the story of how we have understood gravity . . . Chown is excellent on bringing out the temporary nature of theories, as well as the messy business of refining them'

thebookbag.co.uk

'An accessible history of the most well-known but least understood force'

Big Issue North

'Mind-bogglingly brilliant'

www.booklore.co.uk

'Enjoyably, Chown's book doesn't give the sense that "physics is broken" I've come across elsewhere; it's more that we're on the cusp of an exciting step change in our understanding'

Cait MacPhee, professor of biological physics,
University of Edinburgh, *Times Higher Education Supplement*

'A helter-skelter tour through the lives and discoveries of people who helped us understand gravity . . . [Chown] gives us the clearest explanation I have yet read of Einstein's principle of relativity . . . Chown's style carries the reader along in the quest to understand gravity and I recommend it . . . entertaining science history'

John Davies, *Astronomy Now*

'The "detective mystery" aspect of the subject [is] certainly something Chown captures to perfection'

Fortean Times

THE ASCENT OF GRAVITY

The quest to understand the force
that explains everything

MARCUS CHOWN

WEIDENFELD & NICOLSON

First published in Great Britain in 2017
This paperback edition first published in 2018 by Weidenfeld & Nicolson
an imprint of The Orion Publishing Group Ltd
Carmelite House, 50 Victoria Embankment
London EC4Y 0DZ

An Hachette UK Company

1 3 5 7 9 10 8 6 4 2

A CIP catalogue record for this book is
available from the British Library.

ISBN 978 1 4746 0188 7

Typeset by Input Data Services Ltd, Somerset

Printed and bound by CPI Group (UK) Ltd, Croydon, CR0 4YY

To Mike & Claire, Val & Pat, Maureen & Pete
With love, Marcus

It's embarrassing that we're in the twenty-first century and we don't even know what makes gravity work.

Woody Norris

Contents

PART TWO: EINSTEIN

PART THREE: BEYOND EINSTEIN

Acknowledgements

My thanks to the following people who helped me directly, inspired me or simply encouraged me during the writing of this book: Karen, Bea Hemming, Felicity Bryan, Paul Murphy, Michele Topham, Manjit Kumar, Thomas Levenson, David Tong, Andy Hamilton, Lee Smolin, Nima Arkani-Hamed, John English, Tash Aw, Brian Clegg, Graham Farmelo, David Berman, Gennady Gorelik, Neil Turok, Neil Belton, Brian May, Julia Bateson, Nick Booth, Jonathon Tullett, Daniel Tullett, Jose Tate, Barbara Brighton-Pell, Patrick O'Halloran, Sue Noyce, Graham Noyce, Brian Chilver, Pat Chilver, Jean Dyke, Amanda Capewell, Sam Capewell, Grace Capewell, Rob Capewell.

Foreword

Six things you may not know about gravity

1

Gravity creates a force of attraction between you and the coins in your pocket and between you and a person passing you on the street

2

It is so weak that, if you hold your hand out, the gravity of the whole Earth cannot overcome the strength of your muscles

3

Despite its weakness, gravity is so irresistible on the large scale that it controls the evolution and fate of the entire Universe

4

Everyone thinks it sucks but in most of the Universe it blows

5

If it had not 'switched on' after the big bang time would not have a direction

6

Only by figuring it out will we be able to answer the biggest question of all: Where did the Universe come from?

At Livingston in Louisiana and Hanford in Washington State there are 4-kilometre-long rulers made of laser light. At 05.51 Eastern Daylight Time on 14 September 2015, a shudder went through first the Livingston ruler, then 6.9 milliseconds later, the one at Hanford. It was the unmistakable calling card of a passing gravitational wave – a ripple in the very fabric of space-time – predicted to exist by Einstein almost exactly 100 years ago.

In a galaxy far, far away, at a time when the Earth hosted nothing bigger than a simple bacterium, two monster black holes, locked in a death-spiral, swung around each other one last time. As they kissed and coalesced, three whole solar masses vanished, reappearing instantly as a tsunami of a warped space-time, which raced outwards at the speed of light. For an instant its power was fifty times greater than that of all the stars in the Universe put together.

The detection of gravitational waves by the twin detectors of the Laser Interferometer Gravitational-Wave Observatory (LIGO) on 14 September 2015 was an epoch-making moment in the history of science. Imagine you have been deaf since birth, then, suddenly, overnight, you are able to hear. This is the way it is for physicists and astronomers. For all of history we have been able to 'see' the Universe. Now, at last, we can 'hear' it. Gravitational waves are the voice of space. It is not too much of an exaggeration to say that their detection is the most important development in astronomy since the invention of the telescope in 1608.

Gravitational waves confirm that space-time is a 'thing' in its own right, which can quiver and shudder, sending undulations propagating outwards like ripples spreading on a pond. They are the ultimate proof of Einstein's contention that gravity *is* warped space-time. Whereas Newton imagined a 'force' of gravity reaching out from the Sun and ensnaring the Earth like a piece of invisible elastic, Einstein recognised that the Sun creates a valley in space-time in its vicinity around which the Earth circles endlessly like a planet-sized roulette ball in an oversized roulette wheel.

Although Newton's theory of gravity was hugely successful, explaining the motion of the planets and the ocean tides and even predicting the existence of an unknown world – Neptune – Einstein's theory of gravity was just as successful, explaining the anomalous motion of Mercury and predicting the existence of black holes and the big bang in which the Universe was born. But Einstein's theory of gravity, like Newton's before it, contains the seeds of its own destruction. At the hearts of black holes and at the birth of the Universe, it predicts the existence of nonsensical 'singularities' where the parameters of physics skyrocket to infinity.

The irony is that the first force to be described by science and the one everyone thinks was understood long ago is actually the least understood. Gravity, to steal the words of Winston Churchill, is 'a riddle, wrapped in a mystery, inside an enigma'.

Now, at the outset of the twenty-first century, we stand on the verge of a new revolution. The search for a deeper theory than Einstein's – a quantum theory of gravity – is the greatest endeavour ever embarked upon by physics. Already, there are tantalising glimpses of a new world view. Perhaps another Newton or Einstein is at this moment waiting in the wings, assembling the fragmentary pieces of the puzzle into a coherent whole. Or perhaps – a more likely scenario – it will take the efforts of dozens of people working in concert. Many physicists believe we are on the verge of a seismic shift in our view of reality, one more far-reaching in its consequences than any that has gone before.

Will the deeper theory than Einstein's give us warp drives and time machines, the ability to manipulate space and access parallel universes? No one can predict, just as no one in the pre-electrical era could have predicted televisions and mobile phones and the World Wide Web. What we do know is that when at last we have the elusive theory in our possession, we will be able to answer the biggest scientific questions of all. What is space? What is time? What is the Universe? And where did it come from?

But I am getting ahead of myself. How did we get to where we are today, standing on the brink of a vast undiscovered landscape of physics? The story began with a twenty-two-year-old named Isaac Newton in the plague year of 1666 . . .

Author's note

A word on endnotes, which readers will find after the final chapter of the book: some contain asides that, if included in the text, would have broken its flow. Some amplify the explanations in the text, occasionally using technical language. And some are references to books and articles, where you can find out more information about the subject in the text.

PART ONE

Newton

The Moon is falling

*How Newton found the first universal law –
one that applies in all places and at all times*

> For in those days I was in the prime of my age for invention
> and minded Mathematicks & Philosophy more than at any
> time since.
>
> Isaac Newton[1]

> You fainted and I caught you. It was the first time I'd sup-
> ported a human. You had such heavy bones. I put myself
> between you and gravity. Impossible.
>
> Elizabeth Knox, *The Vintner's Luck*[2]

'So, Mr Newton, how did the idea of universal gravity come to
you?'

They are in the garden of Woolsthorpe Manor, half a century after the event: the elderly natural philosopher, now the
most famous personage of his day, sitting across the table from
William Stukeley, the young clergyman and archaeologist who
has set himself the formidable task of writing the first biography
of Isaac Newton. A stream burbles at the bottom of the garden
and lambs bleat at random intervals in the field beyond. A raven
lands on the lush orchard grass before them, pecks at nothing in
particular and takes wing again.

The old man ponders the question, sweeps his long white hair
back from his face, then says: 'Mr Stukeley, you see that tree
yonder?'

'I do.'

'In the spring of 1666, on a warm day not unlike this, I was seated in this very spot, jotting in my notebook, when an apple fell from the tree . . .'

But great men are apt to concoct their own legends. The story of the apple was indeed told by Newton, close to the end of his life, in the garden of Woolsthorpe Manor, Linconshire. 'After dinner, the weather being warm, we went into the garden and drank tea, under the shade of some apple trees,' wrote Stukeley in *Memoirs of Sir Isaac Newton's Life*, published in 1752. 'He told me, he was just in the same situation as when formerly the notion of gravitation came into his mind. It was occasion'd by the fall of an apple, as he sat in contemplative mood. Why should that apple always descend perpendicularly to the ground, thought he to himself . . .?'[3]

The truth, however, is that Newton never once mentioned the tale of the falling apple in the half century after his discovery of the universal law of gravity. Was it true? Or did Newton, his creative days far behind him and his mind now occupied by his legacy, simply see the potential of the story to burn itself into the popular imagination and ensure his immortality? 'Three apples changed the world,' someone tweeted on the death of Steve Jobs, co-founder of Apple computers. 'Adam's apple, Newton's apple, and Steve's apple.'[4]

Nobody knows what led Newton to make his critical connection between heaven and earth, between the force of gravity pulling on the Moon and the force of gravity pulling on an apple. All we know is that the genesis of Newton's universal law of gravity came at a truly horrific time, described so vividly by Daniel Defoe in *Journal of the Plague Year*.[5]

In August 1665, bubonic plague was raging in London. So great was the dread of contamination that in Cambridge, 55 miles to the north-east, the university was closed. Newton, twenty-two years old, unremarkable, unknown, made the trek, by foot, by horse-drawn cart, back to his family farm in Woolsthorpe. There he remained secluded for eighteen months, during which time he not only discovered the universal law of gravity but changed the face of science.

The special one

Isaac Newton was born on Christmas Day 1643. Despite this auspicious date, the 'special one' was so small at birth, reportedly he would fit in a quart mug, and so weak he was expected to die within days.[6]

Newton was a 'posthumous child'. His father had died three months before his birth. His mother was left with little means of support and, when Newton was three, accepted a proposal of marriage from a wealthy rector, almost twice her age. Because he wanted a wife not a stepson, when she moved to his rectory in a nearby village, she had no choice but to abandon Newton to be brought up by his maternal grandparents. Newton despised his substitute parents and later in his notebook confessed to the sin of 'threatening my father and mother Smith to burne them and the house over them'.

On the death of her husband eight years later, Newton's mother returned to Woolsthorpe Manor, bringing with her a half-brother and two half-sisters for Newton. But by this time Newton's sense of rejection by his mother had stoked in him a blind fury that would never be assuaged.

As heir to the family farm, Newton was prohibited from playing with the 'common' children of the agricultural workers. Forced to make his own entertainment, he cut a lonely figure, lost in his imagination, forever building things and investigating things about the world around him. He constructed model windmills and bridges. He cut sundials in stone and, hour by hour, day by day, season by season, recorded the movement of their shadows.

It was because of Newton's singular ability that, when he was twelve, money was found to send him to Kings School in Grantham. The eight miles to the market town was too far to walk each day so he lodged with a local apothecary. Cut off even from family members, he was further isolated. But he fell under the wing of the headmaster, who had a special interest in mathematics and, recognising Newton's exceptional talent, taught the boy everything he knew.

In 1659, when Newton was sixteen, his mother summoned him home to Woolsthorpe to be a farmer and run the family estate, with its woods and streams, barley fields and grazing sheep. But Newton spent his time gathering herbs and reading books.[7] He built water wheels in the stream while the sheep trampled the neighbouring farmer's barley. He let his pigs trespass on others' land, left the fences in disrepair, and was fined in the manor court on both counts.[8] To everyone's relief, including Newton's, he was returned to school in Grantham the following year.

Newton's uncle on his mother's side was another who recognised Newton's unusual abilities. A rector who had studied for the clergy at Cambridge, he helped the eighteen-year-old find a place at the university in 1661. At the time the institution was situated in little more than a dirty and scruffy village. Newton paid his way as a 'sub-sizar', surviving by waiting on wealthier students, running errands for them and eating their leftovers. His undergraduate studies at Cambridge culminated in a Bachelor of Arts degree, awarded to him in January 1665.

Little is known about Newton's experiences as a student. Like his twentieth-century successor Albert Einstein, he appeared not to have distinguished himself in any way. Nevertheless, he studied mathematics and science with intensity, devouring and absorbing the philosophical work of the Greeks. But, crucially, he was critical of what he read. 'Plato is my friend – Aristotle is my friend,' he wrote in his precious notebook, 'but my greatest friend is truth.'

Voyaging through strange seas of thought alone

In 1665, when Newton settled back into life at Woolsthorpe, it was still summer and the air was abuzz with insects and alive with birdsong. So idyllic was the scene that it must have been difficult to believe that, just 100 miles away in London, people were stumbling and dropping in the streets. They were suffering fever and chills and muscle cramps and aching limbs. They were gasping for breath and sometimes coughing up blood. Their armpits and groins were swollen with black buboes as the plague bacterium multiplied in their lymph glands. Before the outbreak was over, 100,000 souls – a quarter of London's population – would be carried away on carts and dumped unceremoniously in plague pits.[9]

Woolsthorpe Manor was a slightly dilapidated two-storey dwelling with grey limestone walls, nestling amid apple trees and grazing sheep on the side of the valley of the River Witham. Seated at his desk, Newton shut all the horrors of his time from his mind. Perhaps he was able to do it because he was psychopathically detached from human suffering. Or perhaps he knew there was nothing he could do. Why worry about things that cannot be changed? Why agonise about things that are in the hands of the Almighty?

Newton was a pragmatist at heart. And a pragmatic man might use a time of terror as an interlude, as a God-given opportunity to penetrate the mind of the Creator. 'My greatest friend is truth,' Newton had written. At Woolsthorpe, while the horror of plague stalked England, Newton began to seek that truth. 'Voyaging through strange seas of thought alone,' he would become the pre-eminent mathematician in the world.[10] He would discover the laws of optics and colours, the mathematics of 'calculus' and the 'binomial theorem'. But, most significantly of all, he would find the universal law of gravitation.

The moment was now ripe for such a discovery because there was a realistic model of the Earth's place in the cosmos. But this had not always been the case.

Mass is the key

Once, it had been thought that the Earth was the centre of the Universe. The mistake was perfectly understandable. After all, the Sun, the Moon and the stars very definitely appear to circle the Earth.

But there are anomalies.

To the ancients, the five naked-eye planets – Mercury, Venus, Mars, Jupiter and Saturn – could not have stood out more prominently if they pulsed on and off like celestial fireflies. They alone crawl snail-like across the backdrop of fixed stars.[11] And, crucially, the pace at which they crawl is uneven. Watch one, night after night, week after week, and occasionally and unexpectedly, it can reverse its direction, and reverse again, describing a crazy loop in the night sky. How is this possible if planets are merely circling the Earth?

The answer is *it is not*.

To explain the anomalous motion of the planets – which comes from the Greek for 'wanderer' – there was concocted an ingenious and cunning scheme. The Greeks were wedded to the idea that the heavens, unlike the Earth, were a realm of utter perfection. And the perfect figure to their minds was the circle. Perhaps, as a planet circles the Earth, it also moves in a smaller circle about its average position? A circle within a circle, or an 'epicycle'. Since motion around the smaller circle allows a planet to travel briefly backwards in its orbit, this would explain why sometimes we see a planet loop back on itself.

This solution to the puzzle of planetary motion is in fact a big con. With enough circles within circles within circles it is possible to mimic absolutely any motion whatsoever. Not only that but the solution is complex and messy. And a key characteristic of modern scientific explanations is that they are simple and economical.

A better explanation of the peculiar planetary motion was proposed by the Polish astronomer Nicolaus Copernicus in 1543. Say the centre of everything is not the Earth but the Sun, and that all of the planets, including the Earth, actually go around the Sun? In this case, Copernicus pointed out in *On the Revolutions of the Heavenly Spheres*, the motion of planets is easy to explain. As it circles the Sun, the Earth regularly catches up and overtakes a planet like Mars, which is orbiting more slowly in its orbit. From the point of view of the Earth, the planet drops behind, appearing briefly to travel backwards against the fixed stars.[12]

Copernicus's explanation of the motion of the planets came at a cost. There were now two bodies about which other bodies circle – the Sun, which ensnares the planets, including the Earth, and the Earth, which holds onto the Moon. And things got even worse when the Italian scientist Galileo zoomed in on the heavens with his new-fangled astronomical telescope. Not only did he see stars invisible to the naked eye, mountains on the Moon and the phases of Venus but, in 1610, he was amazed to find that Jupiter is orbited by four moons. There are not two bodies acting as centres in the Solar System: there are *at least three*.

Ancient ideas were crumbling. According to the Greeks, the most important factor for understanding our world and the Universe was location. Each of the four 'fundamental elements' – earth, fire, air and water – seeks out its allotted place. And all are related to the Earth, with earth, not surprisingly, desiring to get as close to the centre of the Earth as possible. But, in the new view, there was nothing at all special about location. How could there be when there are at least three locations about which other celestial bodies revolve?

The lesson from observing our Solar System is that massive bodies orbit other massive bodies. Location is not the important thing.[13] Mass is the key.

Nature's lonely hearts club force

The pressing question is: how does one mass enslave another? A clue came from magnetism. Lodestones are naturally magnetised chunks of the mineral magnetite. One lodestone attracts another lodestone with a mysterious 'force' that reaches across the empty space between them. As early as the sixth century BC their unusual properties had been remarked upon by the father of Greek philosophy, Thales of Miletus.

In 1600, the English scientist William Gilbert suggested that magnetism might be the force holding together the Solar System. He demonstrated experimentally that the attraction exerted on a piece of iron by a lodestone is bigger the bigger the mass of the lodestone. He also showed that the attraction is mutual – that is, the force of attraction exerted by a lodestone on a piece of iron is exactly as strong as the force of attraction exerted by the iron on the lodestone.

Others such as Robert Hooke, the man who would become Newton's greatest rival, were much taken by Gilbert's findings. But the Sun is a hot body and lodestones heated until red hot were known to lose their magnetism. Hooke therefore saw magnetism as merely a model for the force that is orchestrating the motion of the bodies of the Solar System. Like magnetism, gravity reaches out from one mass across empty space and grabs another mass. Like magnetism, the force is bigger the bigger the masses involved. And, like magnetism, it is a mutual force.

Gravity pulls masses together. It attempts to break their terrible isolation. It is truly nature's lonely hearts club force.

This was the state of play in the plague year of 1666 as Newton sat deep in thought at his desk at Woolsthorpe Manor and began to ponder the nature of the force between massive bodies. He had no more idea what the force of 'gravity' is than what the magnetic force of a lodestone is. But not knowing what the force is did not hamper him. In the words of the twentieth-century physicist Niels Bohr: 'It is wrong to think that the task of physics is to find out how nature is. Physics concerns what we can say about nature.'

Newton knew this truth instinctively. Just because he did not know what gravity is did not mean he could not ask: how does gravity behave?

Reading the book of nature (Kepler's laws)

The vital clues to gravity's behaviour had been discovered by the German mathematician Johann Kepler. Between 1609 and 1619, he had built on the work of the Danish astronomer Tycho Brahe, famous among other things for having a prosthetic nose made of brass after his real one was sliced off in a duel. Brahe had made precise naked-eye observations of the planets from his observatory on the island of Hven, now part of Sweden. After poring long and hard over Brahe's records, Kepler deduced three laws that govern the behaviour of the planets.

Kepler's first law states that the orbit of a planet is an ellipse, with the Sun at one focus. An ellipse is a very specific closed curve, not simply an oval. It can be drawn by pinning two tacks to a flat surface, stretching a loop of string over them, pulling the loop of string taut with a pencil, and moving the pencil point in a complete circuit around them. The two tacks mark the foci of the ellipse. In mathematical terms, wherever a point is located on the ellipse, the sum of the distances to the two foci is the same.

Kepler's recognition that the orbit of a planet is an ellipse was a decisive and significant break with the past. The Greek conviction that circles are perfect had caused them to impose circles on the cosmos. But nature is a book to be read not a book to be written. Realising this, Kepler, and the scientists who followed him, demonstrated more humility than their Greek predecessors. They studied nature and looked to see what *it was telling them*. And what nature was telling Kepler, through the medium of Brahe's painstaking observations, was that the planets are orbiting the Sun not in circles but in egg-shaped ellipses.

Kepler's second law says that a planet does not go around the Sun at a uniform speed but moves more quickly when it is nearer the Sun and more slowly when it is further from the Sun. Actually, the law is a bit more precise than this. It states that an imaginary line joining a planet to the Sun sweeps out equal areas in equal times. Take, for instance, a time interval of 10 days. Two points on a planet's orbit that are 10 days apart can be joined to the Sun to make a triangle. The area of the triangle is always the same irrespective of whether the planet is close to the Sun in its orbit or far from the Sun. It is impossible not to admire the sheer ingenuity of Kepler in teasing out such an odd law from Brahe's observations.

Newton, ensconced at Woolsthorpe, thought long and hard about Kepler's second law. And thinking long and hard was the secret of his genius. Yes, he could build complex things and carry out complex experiments, and he could do both of these things far better than most. But what truly set him apart from all others was his phenomenal, almost unearthly power of concentration. This was the secret of his success. This was his thing.

Newton took no exercise, indulged in no amusements, and

worked incessantly, often spending eighteen or nineteen hours a day writing.[14] The clockwork of his mind whirred incessantly. Every hour spent not studying he considered an hour lost. While others could hold an abstract problem in their mind's eye for fleeting minutes, Newton could focus on a problem for hours, weeks, whatever it took, until finally, he burned through to its inner core and it yielded its precious secret. 'I keep the subject constantly in mind before me and wait 'til the first dawnings open slowly, by little and little, into full and clear light,' wrote Newton.[15]

Newton applied the laser beam of his intellect to Kepler's second law. And eventually, inevitably, he saw what it was telling him about the force of gravity experienced by a planet. And the thing it was telling him has nothing to do with the strength of that force or the way in which that strength changes with distance from the Sun or any other detail like that. A planet sweeps out equal areas in equal times, Newton realised, on one condition and one condition only: that the force it is experiencing is always directed towards the Sun.[16]

Kepler's third law of planetary motion is subtly different from the first two. Instead of describing the individual orbits of planets, it describes how the orbits of different planets relate to each other. It states that the further a planet is from the Sun, the slower it moves and so the longer it takes to complete an orbit. This is a clear indication that the force of gravity experienced by a planet is weaker the further the planet is from the Sun. But there is more in the law than this. Kepler was a mathematical genius. His third and last law actually says that the square of the orbital periods of the planets goes up in step with the cube of their distances from the Sun. So, for instance, a planet that is 4 (that is, 2^2) times as far from the Sun as another takes 8 (that is, 2^3) times as long to complete an orbit.

Kepler's third law is even more esoteric than his second. But do not get hung up on the detail. The key thing is that it is a precise mathematical relation. And that indicates that the force that gives rise to the law – the force between the Sun and the planets – must also be mathematical. This in itself is a revelation. Evidently, nature obeys mathematics. Or, as Kepler might have seen it, God is a mathematician.[17] So the question Newton, frowning at his desk at Woolsthorpe, asked was: what is the mathematical law of gravity?

Newton was in a unique position to answer this question because he alone had defined the concept of a force, transforming it from a hand-waving, nebulous notion into a thing of rapier-sharp scientific precision. In this Newton was indebted to Galileo, who died a year before Newton was born.

Explaining the book of nature (Newton's laws)

Bodies falling under gravity plummet too fast for their fall to be timed precisely with the primitive methods available to Galileo. But he found an ingenious way to dilute gravity and so brake the motion of falling bodies. He set balls rolling down an inclined plane on a table top. The shallower the slope, the more gravity is diluted and the slower a ball rolls. But – and this was a key observation by Galileo – when a ball reaches the bottom of the slope, it continues rolling *at constant speed* until it falls off the edge of the table.

On the table top, which is flat with no slope, gravity is diluted to zero and there is no force on the ball. So Galileo concluded that *in the absence of a force bodies move at constant speed*.

This result is totally counterintuitive. In the everyday world nothing moves at unvarying speed. Kick a stone along the ground and it quickly comes to rest. But the explanation for this, Newton reasoned, is that the stone is subject to a retarding force – the force of friction with the ground. In the absence of such a force – if the stone, for instance, is kicked across a perfectly slippery, ice-covered pond – it will keep on going.

That the natural motion of a body is to keep coasting explains a puzzle which had stumped people ever since they realised that the stars are not actually turning around the Earth but the Earth is instead spinning. Knowing the size of the Earth and that the Earth turns once every 24 hours, it follows that at the equator the Earth's surface is moving at 1,670 kilometres an hour! Why do people living there not notice it? Why, if a ball is dropped there, does the Earth not rotate under it as it falls so that the ball hits the ground far to the east? The answer is that we and the ball and the air around us were all born into a moving world, and continue to move around with the Earth as it turns because that is what moving things do.

Even today no one knows why the natural motion of a body is to keep on coasting. But Newton latched onto Galileo's extraordinary insight and encapsulated it in the first of his three 'laws of motion'.

Newton's first law says that every body either stays at rest or keeps moving forward at constant speed in a straight line unless compelled to change by an external force. (This should not be confused with the Law of Cat Inertia, which states: 'A cat at rest will tend to remain at rest unless acted upon by some outside force such as the opening of cat food, or a nearby scurrying mouse.')[18] According to Newton, a 'force' is something that budges a body from its natural motion – changing its speed or its direction or both. This idea Newton encapsulated in his second law, which says that a body responds to a force by accelerating – that is, changing its speed – in the direction of the force and by an amount that is inversely related to its mass. In other words, a small mass accelerates more than a big mass in response to a given force.

More precisely, Newton's second law states: 'The rate of change of momentum of a body is equal to the force applied'. Newton defined 'momentum' as the product of a body's 'mass' and its 'velocity', which in turn is defined as its speed in a particular direction. Newton, here, was laying the foundations of 'dynamics', the mathematical theory of motion.

That the natural behaviour of bodies is to move in straight lines at constant speed told Newton everything he needed to know about a planet orbiting the Sun. First, no force is required to push a planet around the Sun. This is fortunate circumstance since, as already mentioned, Newton's interpretation of Kepler's second law is that the force of gravity is directed solely towards the Sun, with no component along the path of a planet. A planet simply keeps moving for no other reason than that is what massive bodies naturally do.[19]

Think what an extraordinary revelation this is. Pretty much everyone who had ever thought about the problem of the motion of the planets imagined that some kind of force is necessary to push them around in their orbits. Some imagined invisible angels flying alongside and blowing the planets along or chivvying them with their beating wings. Kepler envisaged magnetic 'spokes' extending from the Sun and impelling the planets as the Sun turned. The French mathematician René Descartes favoured a solar 'vortex' swirling the planets around like cosmic flotsam. But Newton consigned all of these ideas to the dustbin of history. Kepler's second law, he realised, is definitive proof that no force is driving the planets around in their orbits.

That the natural behaviour of bodies is to move in straight lines also told Newton what the force of gravity holding a planet in orbit around the Sun is doing. It is constantly changing its path from a natural straight line to a circle.

Of course, Newton knew from Kepler's first law that the planets orbit the sun not in circles but in ellipses. But ellipses are more complicated figures than circles and, since the elliptical orbits of the planets are pretty close to circles, Newton felt justified in considering them to a first approximation as circular.

The question Newton asked himself was: what is the force required to a keep a body moving in a circle – that is, what is the force needed to continually bend its path away from its natural straight-line trajectory? The answer had already been obtained by others, including Hooke. But Newton did not know this.

Newton sat down with a piece of parchment and drew a circle of radius, r, with a dot, representing a mass, m, on its circumference. The mass, he assumed, is moving at a velocity, v. Now it was just a matter of geometry to work out the force necessary to continually deflect the mass from a straight-line path. The force turns out to be mass times velocity-squared divided by the radius, or mv^2/r.

The formula for this 'centripetal force' actually encapsulates everyday intuition. Say, you swing a stone tied to the end of a length of string around your head. Common sense says the more massive the stone the harder you will have to pull on the string – that is, the bigger the force you will have to apply – to stop the stone flying off on a tangent, trailing the string behind it. Common sense also says that the faster you swing the stone, the bigger the necessary restraining force. And the shorter the string the harder you will have to pull.[20] Gravity is the invisible string that holds onto the planets and stops them flying off to the stars.

Now, Newton asked the following crucial question: if the centripetal force on a planet is provided by gravity, how must that gravity vary with distance from the Sun in order to yield Kepler's third law – that the square of the orbital period goes up in step with the cube of its distance from the Sun? The answer, he discovered, is that the force must weaken with the square of the distance from the Sun. In other words, if a planet is twice as far away as another, the gravitational force it experiences from the Sun is four times as weak; if it is three times as far away, nine times as weak, and so on.[21]

There was one other place in the heavens where Newton could check this 'inverse-square law' of gravity. Jupiter's four moons – Io, Europa, Ganymede and Callisto – had been observed whirling around the planet ever since Galileo first spotted them from Padua in 1610. Astronomers had measured the relative distances of these 'Galilean' moons from Jupiter and timed how long each takes to complete an orbit. They had discovered that the moons orbit Jupiter exactly as the planets orbit the Sun, with their orbital periods varying with their distances from Jupiter as predicted by Kepler's third law. The hard work had been done for Newton. Kepler's third law is an inevitable consequence of a gravitational force that weakens with distance according to an inverse-square law.[22]

The Moon is falling

Kepler's third law, operating in the lofty realm of the heavens, was far removed from the everyday world of sheep grazing in the fields of Woolsthorpe, of hay-filled wagons bumping and jouncing along rutted tracks, of cocks crowing in the cold grey dawn. But Newton was harbouring a revolutionary, heart-stopping thought. What if the force of gravity at work in the heavens is the same force of gravity at work on Earth? What if – and nobody in the history of the world had thought this thought before – what if there is not one law for the heavenly realm and one law for the everyday world? What if gravity is a *universal force* – that operates between every last mote of matter and every other last mote of matter?

Newton was the ultimate pragmatist. He knew that his insight meant nothing unless he could make it count – unless he could use it to calculate something.

The story of Newton's apple, as already mentioned, may be apocryphal. But the point is that, crucially, Newton realised that the force that pulls an apple towards the Earth is the same one that keeps the Moon trapped in orbit around the Earth.

Such a connection between a falling apple and the Moon is not at all obvious. After all, the Moon does not appear to be falling. Newton's genius was to realise that appearances are deceptive.

Imagine a cannon firing a cannonball horizontally across the ground. After travelling a short distance, the cannonball falls to the earth. Picture a bigger cannon that shoots a cannonball faster. The ball travels further before it hits the ground. Now imagine a truly enormous cannon that fires a cannonball at an enormous speed of 28,080 kilometres an hour. For this cannon-ball the curvature of the Earth is now of critical importance because as fast as the cannonball falls towards the ground the ground underneath the cannonball curves away from it. The ball, though it is perpetually descending towards the ground, never gets any closer. Instead, it orbits round and round the Earth, *falling forever in a circle*. 'The knack of flying,' as Douglas Adams so pointedly observed, 'is learning how to throw yourself at the ground and miss.'[23]

The Moon is falling for ever in a circle. So the apple and the Moon *are* doing the same thing. It is just not obvious they are because the apple has no speed parallel to the ground and so falls vertically whereas the Moon, like an ultra-high-velocity cannonball, does have a speed parallel to the ground and so falls in a circle.

Today, children still ask: why does the Moon not fall down? Why do artificial satellites not fall down? What is keeping them up? The thing they do not realise is that nothing is keeping them up. They are falling down! A common misconception is that astronauts in space are weightless because there is no gravity. In fact, gravity even at the altitude of the International Space Station is about 89 per cent of that on the Earth's surface. The astronauts on board are weightless not because they are beyond gravity but because they are falling.

All Newton had to do in order to prove that gravity is a universal force – operating between all masses, whether in the heavens or on Earth – was to compare the effect of gravity exerted by the Earth on the apple with the effect of gravity exerted by the Earth on the Moon. If he was right, the ratio of the two effects should be explicable by a single force which weakens with distance according to an inverse-square law.

Newton turned his attention to a falling apple. He knew – because people like Galileo had measured it – that in its first second of fall an apple descends 490 centimetres (16 feet). The next question Newton needed to answer was: how far does the Moon fall in 1 second?

Newton knew that the Moon is 384,400 kilometres from the centre of the Earth.[24] This enabled him to calculate the circumference of the Moon's orbit. Since he knew that the Moon travels around this orbit once every 27.3 days, he could calculate the speed of the Moon.

The natural motion of the Moon would be to continue at this speed in a perfectly straight line. But, of course the path of the Moon is continually bent away from this straight line and towards the Earth by the force of the Earth's gravity. It is a matter of geometry to calculate how far in 1 second the Moon falls away from a straight-line path and towards the Earth. When Newton did the calculation, he arrived at a distance of 0.136 centimetres (roughly 1/20 of an inch). So now he knew that the gravity of the Earth at the distance of the Moon is 0.136/490 = ~1/3,600th of that at the surface of the Earth (~ means 'approximately').

The Earth's surface is 6,370 kilometres from the centre of the Earth whereas the Moon, as already mentioned, is 384,400 kilometres from the centre of the Earth.[25] In other words, the Moon is about 60 times further from the centre of the Earth than is the Earth's surface. Notice that $60^2 = 3,600$ – the exact amount by which the gravity at the distance of the Moon is weaker than at the surface of the Earth. Newton had proved that a single force that weakens with the square of distance tugs on both terrestrial apples and celestial bodies. Gravity is indeed a universal force.

It is worth pausing for a moment to consider what this means. It means there is a force between every chunk of matter in the Universe and every other chunk of matter. In other words, there is a force of gravity between you and a person walking past you on the street; between you and the mobile phone in your pocket; even between your left earlobe and the big toe of your right foot. In all these everyday circumstances, the force of gravity is far too weak to have any noticeable effect. But gravity is stronger the more stuff there is. It is cumulative. This is why the gravity of the Earth, with a mass of 5.98 million million million million tonnes, is noticeable, and why it pins our feet to the ground.

Because gravity is a universal force, it tries to pull together a collection of massive particles into the most compact form possible, which is a sphere. This can happen only if the matter can flow like treacle, which in practice requires the body to be squeezed very hard by its own gravity. Since ice is easier to squeeze than rock, the threshold mass is different for bodies made of ice than for bodies made of rock. In our Solar System, all icy bodies bigger than about 600 kilometres across are round whereas all bodies smaller than this are potato-shaped. For bodies made of rock, the threshold is about 400 kilometres.

Ultimately, the shape of a celestial body is determined by the strength of gravity, which crushes matter, and the strength of the electromagnetic force, which makes matter stiff so it can oppose gravity. The electromagnetic force between a proton and an electron in hydrogen, the lightest atom, is about 10^{40} – or 1 followed by 40 zeroes – times bigger than the force of gravity between them. So, for the force of gravity to overwhelm the force of electromagnetism, a huge number of atoms need to be in one location, which is why gravity triumphs only for bodies bigger than 400 to 600 kilometres across.

There is a subtlety here. Gravity certainly grows stronger the more matter there is. And this definitely explains why our feet are pinned to the ground by the mass of the Earth. But gravity is not merely a force that big masses exert on smaller ones. It is a mutual force which massive bodies exert on each other. The Earth exerts a gravitational force on our bodies and our bodies exert an equal gravitational force on the Earth. Despite this, we all of course know that we fall towards the Earth and the Earth does not noticeably fall towards us. But this has nothing to do with gravity and everything to do with inertia, the inherent resistance of massive bodies to any changes in their motion.

Bigger masses have more resistance to being budged – in fact, this is the very definition of mass – and the Earth is enormously more massive than a person so enormously more difficult to budge. The British comedy writer Andy Hamilton nailed a profound truth when he quipped: 'Is that why I am attracted to big women and big women are not attracted to me?'[26] Actually, big women *are* attracted to Hamilton but, because of their larger mass, Hamilton's gravitational effect on them is smaller than their effect on him. Similarly, the Earth *does* fall towards you or an apple but by an imperceptibly small amount. 'When Newton sat in his garden,' says philosopher A. C. Grayling, '[he] saw what no one had seen before: that an apple draws the world to itself, and the Earth the apple, through a mutual force of nature that holds all things, from the planets to the stars, in unifying embrace.'[27]

'Millions saw the apple fall,' said the American financier Bernard Baruch, 'but Newton was the one who asked why.'[28]

Faith in simplicity

It was an extraordinary leap of the imagination to see that the Moon, though it does not appear to be, is falling, and that, furthermore, it is falling because of the very same force that causes an apple to fall from a tree: the gravitational pull of the Earth. The heavens at the time were widely considered to be the domain of angels and of God himself. The Greeks had even imagined them made of an ethereal fifth essence, entirely distinct from the everyday 'elements' of earth, fire, air and water. But Newton saw no distinction between up there and down here. In a world still dominated by religious dogma, he had the courage to bring the heavens down to earth. The same laws that govern the behaviour of bodies on Earth govern the behaviour of bodies in the Universe. There exist universal laws – ones that apply *in all places and at all times*. And Newton, a man living at the very dawn of science, whose father, unable to write, had endorsed his will with an X, had penetrated to the heart of nature and seen one such universal law.

It was the first of the great unifications of science. Later, Charles Darwin would unify the world of humans with the rest of the animal kingdom; James Clerk Maxwell would unify electricity, magnetism and light; and Albert Einstein would unify space, time and gravity. Today's physicists seek the ultimate unification – or what they imagine to be the ultimate unification – of gravity and 'quantum theory', the theory of the microscopic world of atoms and their constituents.

But Newton's law of gravity was not only universal, it was simple. 'Truth is ever to be found in simplicity, and not in the multiplicity and confusion of things,' wrote Newton.[29] Had the law of gravity not been simple, of course, it would never have been possible for a man of the seventeenth century – even a man of Newton's genius – to have found it. Think how lucky this is. The Universe at a fundamental level could easily be governed by complex laws, utterly opaque and impenetrable to the three-pound brain of a jumped-up ape not long descended from the trees onto an East African plain. But it isn't. The Universe is orchestrated by simple laws.

Following Newton's lead, others have sought and found yet more simple universal laws. In fact, the belief that such laws exist is the unacknowledged faith behind physics, the light that guides physicists struggling to penetrate the darkness at the frontier of their field. No one knows why the Universe at a fundamental level is simple just as no one knows why it is mathematical. But it was Newton, 350 years ago, who first showed it is both of these things.[30]

Newton's universal law describes the gravitational force between particles of matter. In fact, as Newton was first to realise, ultimately, this is all there is to the Universe: particles and forces. 'The attractions of gravity, magnetism, and electricity, reach to very sensible distances, and so have been observed,' wrote Newton. 'But there may be others which reach to so small distances as to hitherto escape observations . . . some force, which in immediate contact is exceeding strong, at small distances performs the chemical operations abovementioned, and reaches not far from the particles with any sensible effect.'[31] We now know that the electromagnetic force is responsible for Newton's 'chemical operations' and that there are indeed two other fundamental forces of nature which had 'escaped observations' and are exceedingly strong only at small distances.

The job of physicists, as Newton so presciently recognised, is twofold. First, to find the fundamental forces of nature. And, second, to discover how those fundamental forces, working in concert, have conspired to assemble the fundamental particles of nature into the fantastically rich Universe we see around us, complete with galaxies and stars, planets and moons, trees and people.

Twenty-two years of silence

Newton found his universal law of gravity in 1666. But he did not announce it to the world for twenty-two years. No one knows why, though there are several possibilities. One is that when Newton compared the effect of the Earth's gravity at the Moon's distance with the effect of gravity on the ground, it did not confirm the inverse-square law. His seventeenth-century estimate for the distance between the Earth and the Moon was wrong. By the time he realised and discovered the correct value, he had already moved on to other scientific puzzles.

Another plausible reason why Newton did not publish his law of gravity straight away is that he tacitly assumed that the gravitational pull of the Earth is the same as if all of its mass is concentrated at its centre. Recall that, in deducing the inverse-square law, Newton compared the distance of the Moon *from the centre of the Earth* with the distance of an apple *from the centre of the Earth*.

The essence of Newton's theory of universal gravity is that it is a force acting between every piece of matter and every other piece of matter. That means that the gravitational force exerted on the Moon by the Earth is in fact the gravitational force exerted on the Moon by Mount Everest *plus* the gravitational force exerted on the Moon by the core of the Earth *plus* the gravitational force exerted on the Moon by every last sand grain on every beach bordering every continent on Earth . . . In fact, the gravitational pull on the Moon is the sum total of the pull exerted by all the untold zillions upon zillions of particles of matter that make up the Earth.

Newton believed that that pull is exactly the same as if all the matter of the Earth is concentrated at a single point at the centre. Almost certainly he could not prove it. But, in the words of the twentieth-century physicist Richard Feynman: 'You can know more than you can ever prove.'[32] And Newton always knew more than he could prove.

Newton's powers of intuition were formidable. After hours or days or weeks of concentrating on a problem, he would see the solution before him – its inevitability, its obviousness, its rightness. But it is not enough to know the truth. It is necessary to convince others as well. And that meant sitting at a desk with a quill and parchment and dressing up gut instinct with a plodding step-by-step explanation in the toddler language of mortals: mathematics.

One thing was obvious to Newton. The world is a ball with one identical hemisphere on either side of the line joining the Moon to the centre of the Earth. Because of the symmetry of this situation, the gravitational forces exerted by all the chunks of matter in one hemisphere on all the chunks of matter in the second hemisphere will be exactly countered by the gravitational forces exerted by all the chunks of matter in the second hemisphere on all the chunks of matter in the first hemisphere. They will cancel each other out. Consequently, the force of gravity of the Earth on the Moon will act entirely along the line joining the centre of the Earth to the Moon. This is a start. But it is still quite a way from saying that that pull will act along the line joining the centre of the Earth to the Moon as if the entire mass of the Earth is concentrated at a point at the centre of the Earth. This is the thing Newton saw so clearly in his mind's eye in 1666 but could not prove.

Or maybe he could prove it. But just not in the way anyone else living on Earth in 1666 could possibly understand.

In May 1666, Newton invented 'integral calculus'. He called it his 'inverse method of fluxions'. It is a piece of mathematical magic with which he could add up the contributions from an infinite number of infinitesimally small masses (or an infinite number of infinitesimally small *anythings*). It was the perfect instrument to prove that the gravity exerted by the Earth is the same as if all of its mass were concentrated at a point at its centre. But, since Newton had only just invented integral calculus and told not a soul about it, a calculus-based proof was one that Newton, and Newton alone, would understand.[33] Telling the world 'I've got a brilliant proof but, before you can appreciate it, I need to teach you an entire field of abstruse mathematics I've just invented' was unlikely to impress anyone.

But Newton was a complex, and contradictory, beast. In addition to the scientific reasons for not announcing his universal law of gravity in 1666 there may well have been powerful psychological reasons. For a start, he was ridiculously, insanely secretive. At school at Grantham, he was tormented, perhaps for his differentness, by the school bully. According to Newton's own account, after the boy kicked him in the stomach, Newton dragged him to the church by his ears and rubbed his nose against the church wall.[34] Despite his victory, the traumatic experience made

Newton paranoid about exposing any part of himself – even the abstract intellectual constructs of his mind – to the remotest possibility of attack. Possessed of one skin too few, Newton failed to see the robust scepticism of others as an essential part of healthy scientific discourse but instead regarded it as a personal assault by scientific pygmies on ideas he did not need to defend because he knew they were true.

Newton was a prickly, bad-tempered and, at times, vindictive man, who during his lifetime engaged in long-running, bitter and often demeaning feuds with other scientists. There is a strong element of the pot calling the kettle black in Newton's observation: 'We build too many walls and not enough bridges.' There is a certain irony in his statement: 'I can calculate the motion of heavenly bodies, but not the madness of people.'

'Tact is the art of making a point without making an enemy,' said Newton. Unfortunately, he was never able to practise what he preached. He had insight but little ability to act on that insight.

But, then, no person is devoid of all contradictions. The twentieth-century physicist George Gamow told a story about Newton which may or may not be true.[35] Newton loved his cat, said Gamow. And, to let his cat in and out of his study, he cut a hole in his study door – a kind of seventeenth-century cat flap without the flap. But then his cat had kittens. So what did Newton do? According to Gamow, he cut a whole row of little holes in the door . . . *one for each of the kittens*. He was the greatest super-genius of all time and yet he did not realise that the kittens could all go through the big hole.

Newton's obsessive secretiveness may have stemmed from something even deeper. Despite being born a premature weakling, he lived to the grand old age of eighty-four, retaining perfect eyesight and all but one of his adult teeth.[36] When he died, he left behind a box of papers for posterity. So explosive was its content that a bishop who opened it and scanned the documents promptly slammed the lid shut in horror.[37] Among other things, the box contained Newton's writings on religion. He was a deeply religious man who believed in one and only one God. He totally rejected the religious orthodoxy of the 'Trinity' of

God, the Son, Jesus, and the Holy Ghost. His own investigations had revealed that the ideology of the 'three persons in one' Godhead was foisted on the Church by devious means at the First Council of Nicaea, convened in AD 325 by the Roman Emperor Constantine I.

Newton knew his heretical 'Unitarian' belief was enough to make him an outcast. In fact, there were laws in England which explicitly banned anyone with Newton's belief from holding any office of importance and which could even have seen him imprisoned. Newton was a fellow of Trinity College, Cambridge, so no one – absolutely no one – was ever to suspect just how much he abhorred the founding principles of that institution. Newton may have learnt to live a secret life because, in a world where religious beliefs were rigidly and ruthlessly enforced, his life depended on it. And that secrecy, perhaps, had seeped into every crevice, every last cranny, of his life.

So Newton, pacing the rutted lanes around Woolsthorpe, wandering the fields and paths, discovered remarkable things about the world but shared them with no one. He never punched the air and shouted 'Eureka!' but kept his discoveries to himself.

Of course, it is possible to build speculation on speculation about why Newton chose not to share his discoveries in 1666. The fact remains that Newton did not publish his universal theory of gravity for twenty years. What changed everything was a visit to Cambridge by his friend Edmond Halley in August 1684 and a momentous question that Halley asked.

Further reading

Ackroyd, Peter, *Newton*, Vintage, London, 2007.

Feynman, Richard, Leighton, Robert and Sands, Matthew, *The Feynman Lectures in Physics, Volume I*, Addison-Wesley, Boston, 1989.

Gleick, James, *Isaac Newton*, HarperCollins, London, 2004.

Goodstein, David and Goodstein, Judith, *Feynman's Lost Lecture: The Motion of the Planets around the Sun*, Jonathan Cape, London, 1996.

Gott, Richard and Vanderbei, Robert, *Sizing up the Universe*, National Geographic, Washington DC, 2010.

Pask, Colin, *Magnificent Principia*, Prometheus Books, New York, 2013.

Shu, Frank, *The Physical Universe*, University Science Books, Mill Valley, 1982.

The last of the magicians

*How Newton created a system of the world and
found the key to understanding the Universe*

Newton was the greatest genius who ever lived and the most
fortunate; for we cannot find more than once a system of
the world to establish.

Joseph Louis Lagrange[1]

Newton was able to combine prodigious mental faculty
with credulities and delusions that would disgrace a rabbit.

George Bernard Shaw[2]

Edmond Halley was a great fan of Newton's. He might even be
described as his friend – though, as far as personal relationships
were concerned, Newton appeared almost autistic.[3] His meeting
with Newton had come about because of an argument he had
got into in a London coffee house with two friends. One was
Robert Hooke, the man who coined the term 'cell' for the tiny
compartments he had observed stacked together in the tissue of
plants. The other was Christopher Wren, the architect working
on the construction of St Paul's cathedral after the destruction
of its medieval predecessor in the Great Fire of 1666.

Halley had puzzled long and hard over Kepler's third law
and its peculiar diktat that the square of the time taken for a
planet to orbit the Sun is related to the cube of its distance from
the Sun. He had deduced, like Newton, that this could be true
only if the planets are responding to an inverse-square law of
force. Wren and Hooke, sipping their steaming black coffee and

blowing wreaths of smoke from their clay pipes, claimed also to have guessed the inverse-square law. In fact, Wren even asserted that he had known of it many years before Hooke. Hooke, not to be outdone, boasted that, by using an inverse-square law of force, he could explain every last feature of planetary motion. But, when Halley and Wren challenged him to reveal the details, he maintained he was keeping them a secret. Only when more people had tried to do the same as him and failed would he reveal to the world his *tour de force*.

Halley was convinced it was all bluster, a childish game of one-upmanship. As he got up to leave, his friends still arguing, he knew what he must do. Only one man was likely to settle the argument with Wren and Hooke. And that was why, in August 1684, he had taken a hot and uncomfortable coach ride from London to Cambridge.

By now, Newton had a formidable reputation. He had held a permanent chair at the University since 1669. He had been a fellow of the newly founded Royal Society of London since 1672. One year before that he had even delivered to the Fellows of the Society a revolutionary new 'reflecting' telescope. By focusing light with a concave mirror rather than lenses, it created images without the shimmering rainbow colours that plagued all 're-fracting' telescopes.[1]

Newton lodged on the first floor of Trinity College between the Great Gate and the Chapel. In his stuffy rooms with the latticed windows flung wide open, Halley looked down on Newton's large and spacious garden. Hemmed in on all sides by high stone walls, it was accessible only via a staircase inside a wooden loggia projecting from Newton's rooms. The grass in the garden was neatly trimmed. Newton, with his desire for order and perfection, could not bear the sight of a single weed. There was a mature apple tree, a water pump against a wall, and at one end a wooden shed, where Halley knew fires often burned night and day while Newton carried out his most secretive alchemical experiments.

Halley turned to the peculiar, unfathomable man, sitting on the couch, waiting expectantly to hear why his visitor had journeyed all the way from London. He cleared his throat and asked his question: 'Supposing the force of attraction towards the Sun to be reciprocal to the square of their distance from it, what would be the curve described by the planets?'[5]

Newton, without hesitation, replied: 'Why, an ellipse, of course.'

Halley was stunned. He asked Newton how he knew.

'I have calculated it,' replied Newton.

But, though Newton searched among his notebooks and tee-tering piles of papers, he failed to find any sign of the proof. He promised Halley to redo the calculation and send it to him in London.

Newton was as good as his word. Several months later, Halley in London received a proof entitled 'On the motion of bodies in orbit'. In nine short pages of definitions, equations and geo-metrical drawings, Newton demonstrated that the path of a body experiencing an inverse-square law is an ellipse, as stated by Kepler's first law of planetary motion. In fact, he showed that an inverse-square law of gravity, together with some basic princi-ples of motion, accounts not only for Kepler's first law but for *all* of Kepler's laws. Actually, Newton went even further than this. He showed that Kepler's first law in fact describes only a special instance of a body moving under the influence of an inverse-square law of attraction. In general, the path is not an ellipse but a 'conic section'.

Picture a cone standing on its base and a sharp knife that can slice clean through the cone. If the knife simply slices through the cone from one side to the other, the cross-section exposed is elliptical. If the knife cuts down through one side of the cone and out through the base, parallel to the other side, the exposed cross-section is an open-ended 'parabola'. And if the knife cuts down through one side of the cone and out through the base, vertically, the result is an open-ended 'hyperbola'.

The three types of path correspond to three different physical situations. If a body experiencing an inverse-square law force has insufficient speed – or energy – to escape the Sun, it will travel for ever in an ellipse around the Sun. If it has sufficient energy to escape, on the other hand, it will follow a hyperbola, flying off to the stars and never coming back. The parabola is the path of a body that sits on the knife-edge between being bound and unbound. It can escape the Sun's gravitational tyranny only when it has put an infinite amount of distance between it and the Sun, which, in practice, would take an infinite amount of time.

Newton's achievement was monumental. He formulated three laws of motion of a radically differ kind to Kepler's. Although Kepler's are brilliant and precise, they are no more than mathematical descriptions of the manner in which the planets move about the Sun. They are not explanations of why the planets move in the way they do. Newton's laws, on the other hand, describe the motion of all masses, from cannonballs to carriages to planets. They are assumptions about the innermost nature of reality: the relations between matter, forces and motion. And with these three laws, supplemented by his law of gravity, Newton had explained Kepler's second and third laws. He had also taken his laws of motion plus his inverse-square law of gravity and explained Kepler's first law – that the planets travel in ellipses. And he had done it in the tedious toddler language of geometry, understandable by his contemporaries, rather than in the few lines of formulae necessary if he had used his mathematical invention of calculus.[6]

'Newton's demonstration of the law of ellipses is a watershed that separates the ancient world from the modern world,' says physicist David Goodstein of the California Institute of Technology in Pasadena. 'It is one of the crowning achievements of the human mind, comparable to Beethoven's symphonies, or Shakespeare's plays, or Michelangelo's Sistine Chapel.'[7]

The *Principia* – taming the Universe

When Halley finished reading the nine-page treatise Newton had sent him, he was astounded. In his hands, he knew he held nothing less than the key to understanding the Universe.

Immediately, he wrote back to Newton, urging to him to allow Halley to arrange the publication of his treatise. But Newton, the perfectionist, said no. He was not satisfied with his work. He was sure he could improve on it, extend it. He had lots more to say about his laws of motion and his law of gravity and, most importantly, their consequences in the world.

But the dam had finally burst. Halley had created the breach. Newton, for so long the jealous guardian of his discoveries, was now willing to pour them forth. He embarked on a frenetic eighteen-month period in which he honed his ideas and presented them in a form so convincing and so inevitable that no one would for a moment doubt them. The result was the *Philosophiæ Naturalis Principia Mathematica* – 'The Mathematical Principles of Natural Philosophy'. Published on 5 July 1687, the three volumes and 550 pages of the *Principia* not only made Newton's name but they presented an all-encompassing, Universe-explaining 'system of the world'.

Newton's achievement in distilling from the bewildering complexity of the world simple fundamental laws cannot be over-estimated. Today, we think in terms of 'force' and 'mass' and 'velocity'. But someone had to create that vocabulary, invent that framework of thought. And that person was Newton.

He struggled with the chaos of contemporary language, zeroing in on fundamental concepts, giving them knife-sharp definitions above and beyond the slippery vagueness of their everyday usage: 'Absolute space, in its own nature, without relation to anything external, remains always similar and immovable. Absolute, true and mathematical time, of itself, and from its own nature, flows equably without relation to anything external.'[8] It was a titanic struggle, like wrestling a bank of fog to the ground. He was taming the Universe.

According to Pakistani-born Nobel prizewinner Abdus Salam:

Three centuries ago, around the year 1660, two of the greatest monuments of modern history were erected, one in the West and one in the East: St Paul's Cathedral in London and the Taj Mahal in Agra. Between them, the two symbolise, perhaps better than words can describe, the comparative level of architectural technology, the comparative level of craftsmanship and the comparative level of affluence and sophistication the two cultures had attained at that epoch of history. But about the same time there was also created – and this time only in the West – a third monument, a monument still greater in its eventual import for humanity. This was Newton's *Principia*.[9]

Halley himself used the ideas in Newton's *Principia* to show that comets sighted in 1456, 1531, 1607 and 1682 were one and the same body. Travelling in a highly elongated elliptical orbit that takes it far from the Sun, it returns to the inner Solar System and the vicinity of the Earth once every seventy-six years. Halley predicted, correctly, that the comet would return to Earth's skies in 1758. Although he was not alive to see his triumph – not to mention the triumph of Newtonian science – the comet was henceforth known as Halley's Comet.

What is so remarkable about the *Principia* is that here is a man of the seventeenth century discovering deep truth after deep truth about the world with unerring accuracy. 'Nature to him was an open book, whose letters he could read without effort,' said Einstein. Or as Alexander Pope put it: 'Nature and Nature's laws lay hid in night: God said: "Let Newton be!" and all was light.'

Newton himself was more modest about his achievements: 'I do not know what I may appear to the world but to myself I seem to have been only like a boy playing on the seashore, and diverting myself in now and then finding a smoother pebble or prettier shell than ordinary, whilst the great ocean of truth lay all undiscovered before me.'[10]

Despite Newton's humility, the *Principia* is an extraordinary achievement. Three compact volumes that have enabled humans to cross space and set foot on another world, to send space probes sailing out towards the stars, and to understand the motion of distant galaxies turning ponderously in the night.

The last of the magicians

The *Principia* marks Newton out as the pre-eminent thinker of the Age of Enlightenment. This is an extraordinary thing, given his life, because it turns out that science was but one of his interests. In the box Newton left on his death – the one which contained his heretical writings on the Trinity – are other documents. They contained hundreds of thousands of words on his experiments and his thoughts on alchemy, and on his Biblical studies – calculations of the dimensions of the Temple of Solomon, and so on.

Newton was an alchemist, using skills first learnt from the apothecary he boarded with in Grantham to try and transmute lead into gold by recreating the experiments of old. He was also a Biblical scholar, trying to rediscover the wisdom of the ancients. He believed that the Creator had left clues everywhere for him to read. And those clues were not merely scientific ones.

Newton was on a personal quest to understand the world – yet another reason why he felt no urge to share his discoveries with others and why they had to be dragged out of him by a man like Halley. 'To know something that no one else in the world knew or understood – that was a most exhilarating experience of power,' says novelist and historian Peter Ackroyd. 'Perhaps he wished to prolong it for as long as possible.'[11]

Science, alchemy, the Bible, all these things were to Newton equally legitimate ways to understand the Creator's creation, equivalent avenues to God. In fact, Newton spent more time working on alchemy and decoding the Bible than he ever did on science – he even predicted that the world would end in 2060. And this is not to mention the twenty-eight years he spent standardising England's coinage and pursuing counterfeiters as head of the Royal Mint in London.

But if Newton was a man of contradictions it was probably because of his location in history, as author James Gleick points out:

He was born into a world of darkness, obscurity and magic. His name betokens a system of the world. But for Newton himself there was no completeness, only a questing – dynamic, protean, unfinished. He never fully detached matter and space from God. He never purged occult, hidden, mystical qualities from his vision of nature. He sought order and believed in order but never averted his eyes from the chaos. He of all people was no Newtonian.[12]

The twentieth-century economist John Maynard Keynes said something similar. On the 200th anniversary of Newton's birth, he wrote: 'He was the last great mind which looked out on the visible and intellectual world with the same eyes as those who began to build our intellectual inheritance rather less than 10,000 years ago. He was not the first of the age of reason. He was the last of the magicians.'

Further reading

Ackroyd, Peter, *Newton*, Vintage, London, 2007.
Gleick, James, *Isaac Newton*, HarperCollins, London, 2004.
Pask, Colin, *Magnificent Principia*, Prometheus Books, New York, 2013.

Beware the tides of March

*How Newton's theory of gravity is rich in
consequences and can explain not only the motion
of the planets but also the tides in the oceans*

There is a tide in the affairs of men, which taken at the
flood, leads on to fortune. Omitted, all the voyage of their
life is bound in shallows and in miseries.

William Shakespeare, *Julius Caesar*[1]

Time and tide wait for no man.

Proverb[2]

It is a bright and frosty morning in mid-March and a washed-out
Moon, close to full, is hanging in the blue sky. We are waiting,
expectantly, on the river bank, hundreds of us. There is even a
TV crew with a young woman in a red puffa jacket and Burberry
scarf talking to camera. Now and then people glance down at
their watches, then back downstream. But there is nothing to see
except a wide river rolling languorously down to the ocean and a
pair of comedy swans, repeatedly upending their white bottoms
by the opposite bank.

The scene is so tranquil here on the River Severn at Minster-
worth in Gloucestershire that it is impossible to believe that anything out of the ordinary is going to happen. Could it be that we
have driven to this location in the West of England and parked
in this field for absolutely nothing? Could it be that we are all
deluded, gullible victims of some ridiculous hoax?

But then we hear it – a faint rumble like distant thunder.

The swans, startled, right themselves and look all about. The TV reporter in the red puffa jacket breaks off in mid-sentence and swivels to look downstream. And there, suddenly, we see it, preceded by spray, spurting high into the air at a sharp bend in the river bank: a boiling, churning wall of froth and chocolate-brown water spanning the entire 90-metre width of the river and carrying with it kayakers and wet-suited surfers who have been riding the wave all the way from the Severn estuary (the world record, by the way, held by a surfer called Steve Ling, is 14.9 kilometres). Behold the Severn Bore, an angry, stirred-up metre-high hummock of water, racing at up to 21 kilometres an hour the wrong way up the River Severn.

Just as quickly as it arrives, it has gone, disappearing from view around the next bend in the river, heading for the city of Gloucester where it will be truncated by the city's docks. Most but not all of its human cargo have gone with it. Two surfers, who collided as they threaded their boards back and forth across the wave front, are now bobbing up and down in the gently undulating water of the bore's wake, along with the bemused swans.

The TV crew packs up its equipment into shoulder bags and boxes, while the rest of us head back to our cars. Everyone is laughing, light-headed, exhilarated. No one is in the slightest doubt that what they have just witnessed is one of the wonders of the natural world.

Interesting bores

The Severn Bore is one of about sixty bores around the world.[3] The biggest and most terrifying by far is on the Ch'ient'ang'kian River in China. In spring, a monster wave as high as a three-storey house surges upriver faster than most people can run.[4] So great is its roar that it can be heard from 22 kilometres away. Boats must be lifted clear of the river lest they are smashed to matchwood. And every year, despite abundant warning signs erected by the authorities along the riverbank, some people stand too close and are swept away and drowned.

The necessary conditions for a bore are a river estuary of a very particular shape and a large tidal range. The Severn estuary, where the water rises as much as 15.4 metres between low and high tide, has the second highest tidal range in the world. The fast-rising water is funnelled into a channel which rapidly becomes narrower and shallower. Eventually, the speed of the water flowing upriver exceeds that of the water flowing downriver, and a step in water height, technically known as a 'hydraulic jump', is born and travels rapidly upstream. (A similar phenomenon, though static, can be seen in a kitchen sink when water from a tap strikes a basin and spreads out, creating an abrupt change in water height where its speed matches that of the incoming water.) Just as a tsunami is imperceptible out at sea but amplified when it enters shallow coastal waters, the Severn Bore is an imperceptible ripple out in the estuary but grows and picks up speed as it is funnelled by the ever-shrinking channel.

The biggest bores occur in spring and autumn. This is because the Severn Bore and its cousins around the world are merely extreme manifestations of the ocean tides, which are at their largest in spring and autumn. Since the tides are the result of the influence of the Moon it follows that so too is the Severn Bore. Remarkably, a super-localised hummock of speeding water, capable of startling swans and delighting surfers and kayakers, owes its existence to an astronomical body 384,000 kilometres away across space.

The Moon appears so small in the sky it can be covered by a thumb held at arm's length. That it should be orchestrating such a down-to-earth event on a cold March day on the River Severn seems utterly preposterous. No wonder nobody guessed the cause of the Severn Bore. No wonder nobody guessed the cause of the ocean tides. Not for a long, long time.

Baffled by the tides

Nobody knows when the tides were first noticed. But our ancestors left the cradle of Africa and spread across the world on several occasions, beginning 1.8 million years ago with *Homo erectus* and finishing 60,000 years ago with modern humans. Very probably, they made their way around the globe by following the shoreline of the oceans, thus avoiding the obstacles of mountains, deserts and forests, and ensuring an ever-present source of food in the adjacent sea.[5] As they padded barefoot along the wet sand, one thing would have been obvious to our not-quite-human ancestors and our fully human ancestors: twice a day, the sea breathes in and out, surging up a sandy beach before slinking back whence it came. From a clifftop or any other place where the coastline is vertical, it would have been clear that this in-and-out motion is actually a consequence of something more fundamental: twice a day, mysteriously, the ocean rises and falls.

Time passed. Immense tracts of time. People invented farming, started to live in cities and began to speculate about the phenomena shaping the world they found themselves in. By a quirk of geography, the ancient civilisations of the West bordered a sea – the Mediterranean – which experiences barely noticeable tides. People remained ignorant of the phenomenon, and this ignorance had severe consequences for Julius Caesar whose invasion of Britain in 55 BC and 54 BC required him to take a Roman fleet outside the Mediterranean:

> It happened that night to be a full Moon, which usually occasions very high tides in that ocean; and that circumstance was unknown to our men. Thus, at the same time, the tide began to fill the ships of war which Caesar had provided to convey over his army, and which he had drawn up on the strand; and the storm began to dash the ships of burden which were riding at anchor against each other.[6]

'Beware the Ides of March', Julius Caesar is warned by a soothsayer before his murder in the play by William Shakespeare. Perhaps if he had been warned 'Beware the tides of March', his invasion fleet would have suffered less heavy damage in the Atlantic. Such a warning should actually have been possible. Although knowledge of the tides was not widespread in Roman times, their key characteristics had been known since about 330 BC when the Greek astronomer and explorer Pytheas sailed from the virtually landlocked Mediterranean all the way to Britain. On emerging for the first time into the vast open expanse of the Atlantic Ocean, Pytheas made a fundamental discovery.[7] The tides are biggest at new Moon – when the Moon is completely unlit by the Sun – and at the full Moon – when the Moon is completely lit by the Sun. Bizarrely, the tides appear to be controlled by the Moon.

Actually, the observation that the highest tides occur when the Moon and Sun are arranged in space so that the Moon is either completely lit or completely unlit by the Sun strongly hints that the Sun also plays a role in the phenomenon, something also realised by Pytheas. The involvement of the Sun is also supported by the fact that the tides are bigger in spring and in autumn, two very particular times in the Earth's annual journey around the Sun.

Knowing the key characteristics of the tides is obviously a very important first step on the road to understanding the cause of the phenomenon. Nevertheless, for almost two millennia after Pytheas, no one came even close to explaining the baffling spectacle.

At the beginning of the eighth century, the Venerable Bede, an English monk and chronicler, noticed that high tide arrives at different times at different ports around the coast of Britain. The implication was that local geography, as well as the influence of the Moon and Sun, plays a role in determining the characteristics of the tides – an observation reinforced by the absence of significant tides in the landlocked Mediterranean and the presence of giant tides in the funnel-shaped estuary of the River Severn.

As for the cause of the tides, Bede, like everyone else, was, well, totally at sea. He speculated that the Moon blew the ocean inland. And when the Moon had moved a bit so that the ocean was subjected to weaker breath, it returned whence it had come. 'It is as if [the ocean] were dragged forwards against its will by certain exhalations of the Moon,' wrote Bede, 'and when her power ceases, it is poured back again into his proper measure.'

The first attempt at a scientific explanation came from an Arab physician and astronomer in the thirteenth century. According to Zakariya al-Qazwini, the tides are caused by the Sun and Moon heating the water of the ocean, which causes it to expand outwards from the point of heating. Though eminently plausible, the idea fails to explain why the Moon and not the Sun plays the dominant role. The tides pulled by the Moon are about twice as big as those pulled by the Sun.

In 1609, Johannes Kepler, very likely influenced by William Gilbert's recent discovery of the Earth's magnetic field, proposed that the tides were caused by the magnetic attraction of the Moon and Sun on the oceans. Galileo was a big admirer of Kepler's, but he was shocked by this 'childish' suggestion. To him, the whole idea that astronomical bodies could reach out across empty space and affect the Earth smacked of the 'occult'. Galileo instead suggested that the tides are caused by the combined effect of the Earth rotating on its axis and orbiting the Sun, motions which he claimed cause the oceans to slosh back and forth.

The truth is that nobody had the slightest chance of discovering the origin of the tides because nobody had the right mathematical tools to do so. Nobody, that is, until Isaac Newton.

Newton alone created a system of the world, which united the Earth and the heavens in one theoretical framework. Newton alone discovered a universal law of gravity. And that law, he realised, had consequences in domains far removed from the realm of the planets orbiting the Sun. Those consequences he explored methodically in his emerging masterwork, the *Principia*. And chief among them was the tides.

Tides: the lunar connection

In estimating the gravitational force exerted by the Earth on the Moon, Newton had assumed that it is the same as if the entire mass of the Earth is concentrated at a single point at its centre. He had even proved this is so with his new-fangled mathematics of integral calculus. But considering the Earth as a point-like mass is merely a good approximation. The Earth in reality, of course, is an extended body. And because it is an extended body, naturally there are parts of the planet that are closer to the Moon than others. The closer parts experience a stronger pull from the Moon than the other parts. Such differences in gravity, Newton realised, have important consequences. And those consequences are most significant for the oceans because water, unlike solid rock, is free to move.

Consider the point on the ocean immediately below the Moon. The gravitational pull on the water at the surface, which is closer to the Moon, is stronger than the pull on the water at the seabed, which is further from the Moon. This difference in gravity, Newton realised, causes the surface water to be pulled away from the seabed so that the ocean bulges up towards the Moon.

But this is not all. Consider the ocean at a point on the opposite side of the Earth to the Moon. Here, the gravitational pull of the Moon on the water at the seabed, which is closer to the Moon, is stronger than on the water at the surface, which is further from the Moon. The difference in gravity causes the water at the seabed to be pulled away from the surface so, once again, the ocean bulges upwards.

According to Newton's reasoning, then, the Moon creates not one but two bulges in the ocean – one at the point in the ocean closest to the Moon and one at the point in the ocean furthest away from the Moon.[8]

The Earth, though, is not static but spins on its axis. This means that the ocean moves through the two bulges every 24 hours. And from the point of view of someone standing on a beach on the edge of the ocean, the water rises and falls twice every 24 hours. Newton had therefore explained what nobody in history had been able to explain: why there are two tides a day. They are nothing more than a consequence of a universal law of gravity which weakens with distance. But, of course, no one knew of such a law before Newton.

Actually, there is a subtlety here, of which Newton was aware. It is not quite true that the tides at any location repeat every 24 hours. They repeat roughly every 25 hours, something actually noticed by Pytheas in 330 BC.

Picture the Moon again. It does not simply hang static in the sky above a single location on the ocean while the Earth turns beneath it. Instead, it circles the Earth in the same direction as the Earth's rotation, taking 27.3 days to make a complete circuit. This means that a point on the ocean directly beneath the Moon will not be directly beneath the Moon again after 24 hours. In the time the Earth has taken to turn once on its axis, the Moon will have moved on its orbit. For the point on the ocean to be directly below the Moon again the Earth must rotate a further 1/27.3 of a complete turn, which takes 1/27.3 of 24 hours, or about 53 minutes. Consequently, two tides are experienced not every 24 hours but every 24 hours and 53 minutes. This is just one of many reasons why predicting the precise times of low and high tide at any location on a coast requires detailed tide tables.

That the Moon rises 53 minutes later each day and the tides are delayed by 53 minutes each day is yet more evidence that the tides are principally caused by the Moon.

But why are the tides so small in the Mediterranean? The answer is part geography and part ocean depth. As the Earth rotates, the two tidal bulges move westward through the oceans. But this means they head towards the Mediterranean from the direction of the Indian Ocean. Unfortunately, there is a brick wall standing in the way: the landmass of the Middle East. Consequently, no ocean bulge makes it into the eastern Mediterranean.

But what about times when the Moon is directly above the Mediterranean? In this case, the Moon *will* create a bulge in the sea. However, it will be small. The reason is that the difference in the Moon's gravity experienced by water at the surface of the ocean and at the seabed depends on the depth of the water. If the ocean is shallow, the difference is small, and so too is the tidal bulge; if the ocean is deep, the difference is big, and with it the tidal bulge. The Mediterranean is relatively shallow. In fact, its average depth is 1.5 kilometres compared with the 3.3 kilometres of the Atlantic. Consequently, tides in the Mediterranean are less than half as impressive as in the Atlantic even when the Moon is hanging directly above the Mediterranean.

It is not often admitted but the twin tidal bulges in the oceans, which are often shown as huge in textbooks and popular science books, are actually ridiculously small. In mid-ocean, the Moon's gravity lifts the water by at most a metre – little more than a ten-millionth of the Earth's radius. But, of course, an ocean has a very large area, and a metre-high bulge spread over a very large area accounts for a lot of water. When that water sloshes into the shallows around the land, it is amplified in height exactly like in a tsunami. Though the tides at mid-ocean are unnotice-able, along the shorelines of the ocean they can be more than ten times as big.

Tides: the solar connection

As Pytheas discovered, the tides are not caused by the gravita-tional pull of the Moon alone but by a combination of the pull of the Moon and the Sun. The reason these two bodies are respon-sible is simple. They are the celestial objects with the dominant gravitational pull on the Earth. The Moon is enormously less massive than the Sun but an awful lot closer – and its closeness wins out. This is why the tides pulled by the Moon are twice as big as those pulled by the Sun – from which it can be deduced that the Moon is twice as dense as the Sun.[9]

The biggest tides occur, as expected, when the effects of the Earth and the Sun reinforce each other. This happens in spring and autumn. It is not easy to visualise. But the key thing is that the Earth spins like a top tipped at 23.5 degrees to the vertical. This means that the Moon's orbit is also tipped.[10] The geometry of the situation means that the only time the Moon and Sun can be perfectly aligned and so pull on the Earth's oceans with maximum force is when the Earth in its orbit is halfway between summer and winter – that is, in spring and autumn.

The perfect alignment also requires the Moon and Sun to be either on the same side of the Earth, so that the Moon is in shadow – a new Moon – or on opposite sides of the Earth, so that the Moon is completely illuminated – a full Moon. This is why the biggest tides – and the biggest Severn bores – occur in spring and autumn around the time there is a full or new Moon in the sky.[11]

The Moon and Sun do not exert a tidal effect only on the oceans; they exert a tidal effect on the whole planet. But, because the rock of the Earth is more rigid than water, the effect on the land is far smaller and much harder to spot. Remarkably, though, tides in the land were first noticed – though not understood – in antiquity.

Tides in the rock: wells and springs

The tides have many baffling features. They occur, after all, twice every 25 hours not every 24 hours. They vary according to the seasons and according to the phases of the Moon. And they vary according to local geography. But one feature of the tides – first noticed by the Greek philosopher Poseidonios – seems more baffling than all the rest.

Poseidonios, who lived between 135 and 51 BC, made observations of the tides in the Atlantic off the coast of Spain. He also observed water in wells. And what he noticed was something very peculiar. As the water in the ocean rises, the water in wells falls, and vice versa. Poseidonios' original observations are lost. But the Greek geographer Strabon, who lived from 63 BC until about AD 25, reports them in his *Geographika*:

> There is a spring at the [temple of] Heracleium at Gades [Cadiz], with a descent of only a few steps to the water (which is good to drink), and the spring behaves inversely to the flux and reflux of the sea, since it falls at the time of the flood-tides and fills up at the time of the ebb-tides.

What in the world could cause the water in such a localised region as a spring or well to do the exact opposite of what the water in the ocean is doing? There was unlikely to be an answer while the cause of the tides remained a mystery. In fact, incredibly, the puzzle was solved only in 1940 by an American geophysicist called Chaim Leib Pekeris.[12]

A tide can be defined as the distortion in the shape of one body caused by the gravitational pull of another body not simply a distortion in the shape of its water. And, in fact, the pull of the Moon causes a tidal bulge in the rock immediately beneath it in exactly the same way that it causes a tidal bulge in the ocean immediately beneath it. The bulge in the rock is a lot smaller on account of rock being a lot more rigid than water. Twice every 25 hours, then, the solid Earth at any location bulges upwards and shrinks back down again, stretching and squeezing the rock.

Now, say the rock into which a well is dug is porous so that it contains water. This is not unlikely since the very fact that a well contains water means there must be water in its vicinity. The surrounding rock is therefore like a waterlogged sponge. And, like a waterlogged sponge, it sucks water out of a well when the rock is stretched and squirts it back into the well when the rock is squeezed.

The rock and ocean are stretched at high tide and squeezed at low tide. Consequently, at high tide, water is sucked out of a well, lowering its water level, and at low tide water is squirted back into a well, raising its water level. This is precisely the phenomenon observed by Poseidonios. It took 2,000 years but Pekeris finally explained it.

Tides in the rock: the LHC

There is a more contemporary and more high-tech example of the effect of the tides on the solid planet. At CERN, the European laboratory for particle physics, near Geneva, subatomic particles are whirled at dizzying speed around a subterranean racetrack 26.7 kilometres in circumference. While cows graze peacefully in fields spanning the border between France and Switzerland, 100 metres or so below them the microscopic building blocks of matter are slammed together in collisions of unimaginable violence. The energy of motion of the incoming particles is converted into the mass-energy of new particles, which are conjured out of the vacuum like rabbits from a hat.[13] As the subatomic shrapnel speeds outwards from the collision point, it is detected by cathedral-sized detectors. It was in such collision debris, for instance, that the Higgs particle (the 'quantum' of the Higgs field, responsible for endowing all other subatomic particles with their masses), was discovered in July 2012.

The Higgs was found with the Large Hadron Collider, which whirls beams of protons both ways around the underground ring at 99.9999991 per cent of the speed of light before slamming them into each other.[14] But the LHC occupies the circular tunnel previously used by another particle accelerator: the Large Electron-Positron Collider, which instead smacked together electrons and their antiparticles, positrons. And it was while using the LEP in 1992 that physicists noticed something peculiar about the energy of the particle beams.[15]

More than 3,000 electromagnets distributed around the circular LEP tunnel constrained the electrons and positrons, continually bending their trajectories away from the straight-line trajectory their inertia wanted them to take. But the LEP physicists noticed that, twice every 25 hours, the beams drifted slightly from their path and back again. In order to keep the beams from wandering outside the giant ring, the physicists had to continually compensate for the drift by slowly increasing the energy of the particles, then reducing it again. The necessary change in energy of the beams was tiny – about a hundredth of a per cent.

What could possibly be causing the particle beams to drift periodically from their circular path? After the physicists had puzzled for a while, they finally realised. The tides rise and fall twice every 25 hours. Incredible as it seemed, the effect observed at the LEP was connected to the tides.

Twice every 25 hours the rock into which the LEP ring was bored bulged upwards. This stretching of the rock caused the LEP to shrink. And twice every 25 hours the crust dropped back down, compressing the rock and expanding the LEP. The crust moved up and down by only 25 centimetres, roughly the height of the book you are reading, and this changed the circumference of the LEP by at most 1 millimetre.[16] Nevertheless, it was enough to require the energy of the circulating particles to be periodically adjusted by about 0.01 per cent lest they wander from the ring.[17]

The effect was of course largest when the Moon was full or

when there was a new Moon – times when the Sun and the Moon are aligned and reinforce each other's effect on the Earth. It is hard to imagine a more high-tech manifestation of the effect of the tides on the solid Earth.[18]

Moonquakes

But the rocks on the Earth are not alone in experiencing tidal stretching and squeezing. So too do the rocks on the Moon. In fact, the tides the Earth pulls on the Moon are much bigger than the tides the Moon pulls on the Earth because the Earth is about 81 times more massive than the Moon. It might be expected that the tides on the Moon are also 81 times bigger than those on the Earth. But, remember, tides are not caused merely by gravity but by differences in gravity. And the Moon is only about a quarter the diameter of the Earth, which means it has only a quarter of the length span over which such a difference in gravity can manifest itself. So the tidal stretching of the Moon by the Earth is not 81 times bigger than the tidal stretching of the Earth by the Moon but only a quarter of that figure, or about 20 times.[19] Nevertheless, it is enough to stretch the Moon by about 10 metres.

We tend to think of the Moon as stone-cold dead, its grey, crater-strewn desolation untouched by the hand of change. But this tidal stretching and squeezing means that the Moon is not quite the inert world of popular imagination. In fact, since well before the invention of the telescope, people have reported seeing strange lights on the Moon at a rate of once every few months. One of the earliest sightings, for instance, was made on 18 June 1178 by five monks at Canterbury Cathedral who reported an explosion on the Moon. The mysterious lights, known as Transient Lunar Phenomena, are one of the greatest mysteries of the Moon.

TLPs that have been observed in the age of the telescope share a number of common features. They are localised, slightly bigger than the resolution limit of the human eye, implying they cover an area of at least 1 square kilometre. They last from a minute to a few hours. They involve a brightening, dimming or even blurring of the lunar surface. And before they disappear they sometimes change colour to a ruby red.

For a long time many astronomers believed that TLPs were 'in the eye of the beholder' and not a phenomenon intrinsic to the Moon. But, in 2002, Arlin Crotts of Columbia University in New York sifted through the records of 1,500 historical sightings. He discovered that most reliable TLPs occur at just six locations on the Moon – half at the 45-kilometre-diameter Aristarchus crater and a quarter at the 100-kilometre Plato crater.[20]

The six locations are all places where the lunar crust has been violently fractured, either by relatively recent asteroid or comet impacts – within the last few hundred millions of years – or by the flurry of mega-impacts which occurred 3.8 billion years ago and caused lava to well out of the Moon's interior and form the lunar 'seas', or Maria.[21]

Seismometers left on the Moon by all but one of the Apollo missions have recorded several hundred 'moonquakes', which, not surprisingly, are more common when the tidal effect of the Earth is greatest. Most have been located along the boundaries of the mare basins where the rock is most fractured. Not only that but Apollo 15, Apollo 16 and the Lunar Prospector probe, which orbited the Moon in 1998, all detected occasional bursts of radioactive radon-222 gas on the surface, and these were associated exclusively with the six TLP sites.

Radon-222 is a decay product of uranium, which is distributed throughout the rocks of the Moon's interior. Crotts consequently speculates that TLPs occur when moonquakes cause gas from deep in the lunar interior to vent through fissures and cracks. The gas often builds up pressure before bursting its plug of lunar soil, or 'regolith', and exploding into space.

Crotts thinks a mere half a tonne of gas escaping into the vacuum would be enough to puncture the regolith, creating a cloud a few kilometres across that persists for between 5 and 10 minutes. The gas cloud either plunges the surface below into shadow or shines brightly because the dust grains it contains reflect more light when scattered through the vacuum than when clumped together on the surface. It is also possible that friction between the grains separates out negative and positive electrical charge, eventually triggering a 'breakdown discharge' like lightning which energises the atoms of the gas, causing then to emit characteristic red light.

According to Crotts' calculations, the periodic tidal stretching and squeezing of the Moon by the Earth's gravity grinds up about 100,000 tonnes of rock a year – a mass equivalent to one aircraft carrier. And from this is released about 100 tonnes of gas.

None of this speculation is academic because there are plans for humans to go back to the Moon. Apollo 18, which was cancelled before its launch, was actually scheduled to land at one of the principal TLP sites. If a TLP happened at a landing site it would be very dangerous for any astronauts. Picture the scene:

20 July 2025, Aristarchus crater, lunar nearside: exactly 56 years after Apollo 11, NASA's Altair 2 landing craft touched down only hours ago and astronauts are now walking on the Moon again for the first time in more than half a century. Suddenly, a large area of the crater floor begins to convulse and a titanic explosion of gas sends dust fountaining up into the vacuum. Knocked off their feet by the blast, the astronauts look back towards their landing craft. But it is no longer there. It has disappeared in a roiling cloud of silver dust.

If Crotts is right, the Moon is a more dangerous place for humans than anyone suspected. And it is entirely a consequence of Newton's theory of tides.

Since moonquakes can be triggered by tides pulled by the Earth in the Moon's rock, it is natural to wonder whether terrestrial earthquakes are triggered by tides pulled by the Moon in the Earth's rock. It seems they are not – at least not the big earthquakes. But, interestingly, the aftershocks of the devastating earthquake which struck Christchurch in New Zealand on 22 February 2011 were found to be correlated with the location of the Moon in the sky.[22] A possible reason for this may be that the big quake left the rock in an unstable state, ready to move again if nudged by even the slightest force.

Tidal spin-down of the Moon

The tides on the Earth and Moon do more than simply distort the shape of each body, causing the rise and fall of the seas on Earth and moonquakes on the Moon. They have profound consequences for the system of the Earth and Moon as a whole. Once upon a time, for instance, the Moon spun faster than it does today. Its rotation was slowed by the tidal interaction with the Earth.

When the Moon was spinning faster, the bulge pulled in the Moon by the Earth was dragged around with the rotation of the Moon so that the bulge no longer quite faced the Earth. The Earth's gravity pulled back on this receding bulge and the effect of this was to brake the rotation of the Moon. Eventually, a point was reached when the Moon was spinning so slowly that it was turning only once on its axis during each orbit of the Earth.

This is the case today. One face of the Moon – the lunar near-side – perpetually points towards the Earth while the lunar far side perpetually faces away from the Earth. In fact, the far side of the Moon was seen for the first time only on 7 October 1959 when the Soviet 'Luna 3' space probe flew over it.[23]

Because of the Moon's 'synchronous' rotation, the tidal bulge caused by the pull of the Earth points perpetually towards the Earth. Since the bulge is no longer being dragged around by the Moon's rotation, the Earth's gravity, which formerly pulled back on the receding bulge, braking the Moon's spin, no longer has any effect on the Moon's rotation. In fact, the Moon's rotation has been 'locked' in this state ever since the moment the Moon's rotation period first matched its orbital period.

Tidal spin-down of the Earth

But the Moon is not alone in having its rotation slowed by a tidal interaction. The rotation of the Earth is also slowed. The effect is less dramatic than in the case of the Moon because the Earth, being a far heavier flywheel, is more resistant to having its motion changed. Consider the bulge created in the ocean on the side of the Earth facing the Moon. Because the Earth is spinning quickly, the bulge tends to get ahead of the line joining the Earth to the Moon.[24] The Moon's gravity pulls back on this receding bulge, braking the Earth's rotation.

The unavoidable conclusion is that the Earth must have spun faster in the past. And evidence supports this. It comes from corals. Such marine organisms, most commonly found in tropical seas, secrete calcium carbonate to form a hard skeleton. The daily and seasonal growth of the skeletons creates regular bands in much the same way that the yearly growth of trees creates tree rings. By counting the bands, it is possible to determine how many days there are in a year. The evidence from fossil corals which lived about 350 million years ago is that at the time there were about 385 days in a year. Since the year – the time taken for the Earth to circle the Sun – is unlikely to have been different, it must mean that the day 350 million years ago was less than 23 hours in length.[25]

A reduction of just over an hour in the day in 350 million years indicates a relatively modest slowing down of the Earth's spin. But the slowdown is remorseless and ongoing. We know, for instance, that the day today – if that makes sense – is longer than the day of a century ago by about 1.7 milliseconds. In fact, we can be sure that for the past 2,500 years the length of the day has been increasing at 1.7 milliseconds per century. The evidence, remarkably, comes from Babylonian clay tablets.[26]

Babylonian astrologers used such tablets to record total eclipses of the Sun, when the disc of the Moon slides across the solar disc, plunging the world into darkness in the middle of the day. Most of the tablets were unearthed in the nineteenth century by peasants looking for bricks and sold to antique collectors in Baghdad, 85 kilometres to the north of the ancient city of Babylon. From there they found their way to the British Museum in London, which boasts an almost complete collection. Many of the clay tablets record the precise times of the total eclipses.

But the timings pose a puzzle.

In 136 BC, for instance, an astrologer recorded that, at 8.45 a.m. on the morning of 15 April, Babylon was plunged into darkness

when the Moon passed in front of the Sun. There is no reason to
doubt the astrologer's account. But, if present-day astronomers
use a computer to wind back the movements of the Earth, Moon
and Sun, like a movie in reverse, they find something puzzling.
The total eclipse of 15 April 136 BC should not have been visible
from Babylon. The Earth, Sun and Moon were simply not lined
up in such a way to create a total eclipse. In fact, the 'zone of
totality' should have passed over the island of Mallorca, 48.8
degrees west of Babylon.

A difference of 48.8 degrees amounts to one-eighth of a complete rotation of the Earth, or 3.25 hours. It seems that, during
the total eclipse of 15 April 136 BC, the Earth was one-eighth of
a turn to the east of where it should have been. There is only
one way to explain this. Over the past millennia, the Earth's spin
must have slowed down. Since 136 BC there have been about a
million days, so, even if the day was only a fraction of a second
longer back then, all those fractions of a second would have
added up to explain the 3.25-hours discrepancy in the timing of
the 136 BC total eclipse. In fact, the only way to make sense of
the Babylonian eclipse records is if the day in 500 BC was about
1/20 of a second shorter than it is today, and that ever since it has
been lengthening by 1.7 milliseconds per century.

It is wonderful that marks scratched onto a clay tablet by an
ancient civilisation can yield such super-precise astronomical information. The phenomenal accuracy of the techniques is down
to the astronomical coincidence that the Moon and Sun appear
the same size in the sky. This leads to an eclipse 'track' which is at
most 250 kilometres wide, making total eclipses at any given spot
on Earth very rare indeed. So if someone in antiquity recorded an
eclipse at a particular location, it is not necessary for astronomers
today to know a precise date in order to identify it. Knowing the
year to within 20 years either way is usually good enough.

There is a twist to this story. Subtle changes in the shapes of
the orbits of artificial satellites caused by the Earth's tidal bulge
imply that tidal braking of the Earth should in fact be lengthening the day by 2.3 milliseconds per century not 1.7 milliseconds.
Something else must be affecting the Earth's spin. The something

else turns out to be connected with the last ice age, which finished about 13,000 years ago.

During the ice age, the tremendous weight of the ice sheets bore down on the Earth, flattening the planet slightly at the poles and making it fatter. At the end of the ice age, when the ice began to melt, the land began slowly to rise. This process of 'post-glacial rebound' is still going on today. Its effect is to make the Earth more circular and less fat. Consequently, like an ice skater pulling in their arms, the planet spins ever faster. The effect causes a shortening of the day by between 0.5 and 0.6 milliseconds per century, and explains why the day is currently lengthening not by 2.3 milliseconds but by only 1.7 milliseconds per century.

In the long term, the braking effect of the Moon on the Earth's rotation might be expected to slow it to the point at which one face of the Earth perpetually faces the Moon just as today one face of the Moon perpetually faces the Earth. If this were to happen, the Moon would be invisible from one half of the planet just as the Earth is today invisible from one half of the Moon. Calculations show that such a 'locking' of the Earth's rotation will occur when the Earth's spin has slowed so much that it turns on its axis once every 47 present-day days.

It will take more than 10 billion years for the Earth's rotation to slow down this much. By that time, the Sun will have run out of hydrogen fuel in its core, swelled into a red giant and either incinerated or swallowed both the Earth and Moon. The truth is that the Earth's rotation will never become locked liked that of the Moon. There is simply not enough time available. Nevertheless, there are systems out in space where this has indeed happened. Stars whirling around each other in close 'binaries' are expected to have their rotations tidally locked, with each star perpetually showing its partner the same face. And, closer to home, Pluto and its moon, Charon, are both tidally locked.

The fleeing Moon

The tidal effect of the Moon on the Earth slows the rotation of the Earth, reducing its 'angular momentum'. There is a fundamental edict of physics known as the 'conservation of angular momentum', which says that the angular momentum of an isolated, or 'closed', system can never change. So, if the angular momentum of the Earth goes down, the angular momentum of something else must go up to exactly compensate for it. That something else can only be the Moon.

The Moon's gravity creates two bulges in the ocean – on opposite sides of the Earth – but the one closest to the Moon is the one with the strongest and most significant gravitational pull on the Moon. As mentioned before, this tidal bulge tends to race ahead of the Moon's orbit because the Earth revolves on its axis in less time than it takes the Moon to go around the Earth. So its gravity tends to drag the Moon along in its orbit, speeding it up.

Remember that the Earth's gravity at the distance of the Moon is exactly that required to bend the trajectory of a body moving at the Moon's speed into the closed orbit we see. So, if the Moon speeds up, it finds itself travelling too fast for its orbital distance and is effectively flung outwards. Outwards – that is, further from the Earth – is 'up', and, as we know, when an object such as a ball is thrown upwards, gravity slows it down. So, paradoxically, the Moon, which is speeded up by its tidal interaction with the Earth, ends up moving more slowly in an orbit further from the Earth. And this does indeed increase its angular momentum as required.[27]

This is not just theoretical. The manned American spacecraft, Apollo 11, 14 and 15, and the unmanned Russian rovers, Lunokhod 1 and 2, all left reflectors on the lunar surface. The fist-sized mirrors, known as 'corner-cubes', have the property that they reflect back light in exactly the direction it comes from. It is possible to fire a laser beam at the Moon, bounce it off a corner-cube reflector, and time how long it takes for the beam to return to Earth. Knowing the speed of light, it is a simple matter to deduce the distance of the Moon.[28]

The experiments reveal that every year the one-way journey of a laser beam gets longer by 3.8 centimetres.[29] In other words, the Moon is receding from the Earth by a thumb's length every 12 months. If you have made it to 70, during your lifetime the Moon will have receded by the length of a family car.

The visibility of total eclipses

That the Moon is moving away from the Earth at 3.8 centimetres a year obviously tells us it was closer in the past. And this has implications for the visibility of total eclipses, one of the most spectacular of natural phenomena.

As mentioned before, a total eclipse occurs when the Moon passes between the Earth and the Sun, blotting out the solar disc and bringing midnight in the middle of the day. Such an event is possible because, although the Sun is about 400 times bigger than the Moon, it is also about 400 times further away. This means the two bodies appear the same size in the sky. This is a very fortunate circumstance for us. Even though there are 170-odd moons in the Solar System, there is not another planetary surface from which such a perfect eclipse can be seen. But we are not simply lucky to be in the right place, we are also lucky to be alive at the right time.

Because the Moon is moving away from the Earth, in the past it appeared bigger and in the future it will appear smaller. It turns out that there were no total eclipses before about 150 million years ago and there will be no more after another 150 million years. Total eclipses have been possible for only a few per cent of the lifespan of the Earth. For most of the reign of the dinosaurs, for instance, there were no total eclipses.

That the Moon is moving away from the Earth and was closer in the past also ties in neatly with an idea about the birth of the Moon.

The planet that stalked the Earth

The Moon is unusually large compared to the Earth – about a quarter of its diameter. All other moons in the Solar System are tiny compared to their parent planets. Granted, Pluto has an even bigger moon relative to its size, but Pluto has not been considered a fully fledged planet since 2006.

The unusual size of the Moon is a hint that it had an unusual origin. In fact, it is believed that shortly after the birth of the

Earth 4.55 billion years ago, the planet was struck by a body with a mass similar to Mars. The titanic collision with 'Theia' liquefied the exterior of the Earth, splashing mantle material into space to form a ring, not unlike the rings seen today around the gas giant planets of our Solar System. The material of the ring congealed quickly into the Moon – but orbiting about ten times closer than it does today. Thereafter, the Moon began moving away from the Earth.

The key evidence for this 'Big Splash' theory of the Moon's origin came from NASA's Apollo programme, which found that the Moon is made of material similar to the Earth's exterior 'mantle'. Its rocks are also drier than the driest terrestrial rocks, indicating that intense heat once drove out all their water. The problem has been that for a Mars-mass object to create the Moon and not shatter the Earth it must have made a glancing blow at a very low speed. But bodies orbiting the Sun, both inside and outside the Earth's orbit, are moving far too fast.

The Big Splash theory can be made to work if Theia actually shared the same orbit as the Earth. This could have happened if it formed from rubble that accumulated at a stable 'Lagrange point', either 60 degrees behind or 60 degrees ahead of the Earth in its orbit around the Sun.[30] Today, similar asteroidal rubble can be seen orbiting the Sun 60 degrees behind and 60 degrees ahead of Jupiter, becalmed in a kind of gravitational Sargasso Sea. According to this twist on the Big Splash theory, Theia stalked the Earth for millions of years before being nudged into a catastrophic colliding orbit.

While the gravity of a body weakens with the inverse-square of the distance from the body, the tidal force, which is due to differences in gravity, weakens as the inverse-cube. So, at the distance from the Earth at which the Moon formed – about ten times closer than it is today – the tidal force it exerted on the Earth was $10^3 = 1,000$ times greater than today. The Earth at the time, being still molten from its fiery birth, is unlikely to have had any oceans. But say it had: twice daily the sea would have risen not by metres *but by kilometres*.

The new-born Moon did not only exert a tidal effect on the

Earth, the Earth exerted a tidal effect on the Moon. And that effect too was 1,000 times bigger than today. In fact, the tidal braking of the Moon's rotation was so huge that it is probable that its rotation became locked very early on – perhaps within only 10 million years of its violent birth. Since the first micro-organisms on Earth appeared much later, probably between 4 and 3.8 billion years ago, no living thing has ever looked up at the night sky and seen the far side of the Moon rotate into view.

The Moon was not always fleeing so fast

An interesting question is: Has the Moon always been receding from the Earth at 3.8 centimetres a year? In 2013, a team led by Matthew Huber of Purdue University in West Lafayette, Indiana, considered the situation 50 million years ago. They fed data on ocean depths and the contours of the continents that existed at the time into a computer model that simulates tides. They con-cluded that, 50 million years ago, the Moon was receding from the Earth at a slower rate, perhaps only half as fast.[31]

The key is the North Atlantic Ocean, which today is wide enough for water to create a large tidal bulge, which can pull on the Moon, causing it to recede relatively quickly. About 50 mil-lion years ago, the Atlantic had not grown to its present size so the tidal bulge created by the Moon in the Atlantic was smaller and its effect on the Moon's recession was less marked. At the time, most of the Earth's tidal effect on the Moon in fact came from the Pacific Ocean.

This is yet another illustration of the complexity of the ocean tides. How big they are and how much they brake the Earth's rotation and the speed of the recession of the Moon depends on how easy it is for tidal bulges to move through the oceans. This in turn depends on the arrangement of the continents, which is continually changing over geological time because of continental drift, more correctly known as plate tectonics.

It is the long-term unpredictability of plate tectonics that makes it hard to predict how long it will take for the Earth's spin to slow sufficiently that it perpetually presents one face to the Moon. It is possible to say only that this state, in which the Earth spins on its axis once every 47 present-day days and the Moon has receded to the point that it takes 47 days to orbit the Earth, will be achieved after more than 10 billion years. But, as pointed out, this is totally hypothetical since the Sun will have evolved into a monstrous red giant 10,000 times as luminous as it is today and destroyed or at least disrupted the Earth–Moon system.

Tides have a final twist. Every day as the sea surges up beaches around the shorelines of the continents, countless pebbles are tumbled and smashed together. Friction between them generates heat-energy, which ends up in the environment. In fact, it is the loss of energy in this way that is ultimately responsible for slowing down the rotation of the Earth.

The tidal heating of the Earth is modest. If you wade into the sea, the sand and pebbles are not likely to scorch your feet. But there is one place in the Solar System where tidal heating is not modest at all: Jupiter's giant moon Io, discovered by Galileo in 1609.

Pizza moon

It is 8 March 1979. NASA's Voyager 1 space probe has streaked through the Jupiter system faster than a speeding bullet. It is now heading towards a rendezvous with Saturn in autumn 1980. But, before the probe leaves the gas giant planet forever, the Voyager team orders its camera to point back the way the space probe has come and take a parting shot of Io. Navigation engineer Linda Morabito is the first to see the image after its 640-million-kilometre journey back to Mission Control. When she does, her heart misses a beat. Spouting from the tiny crescent moon, silhouetted against the starry backdrop of space, is a phosphorescent plume of gas.

Morabito is the first human in history to see the super-volcanoes of Io. Over the next days, as the Voyager team pore over image-enhanced photos and thermal data, they spot a total of eight gigantic plumes, pumping matter hundreds of kilometres into space. Io turns out to be the most geologically active body in the Solar System, with more than 400 volcanoes. The vents that pepper the orange and yellow and brown of its pizza-like surface are reminiscent of the geysers of Yellowstone Park. In fact, strictly speaking, that is what they are: geysers not volcanoes. Lava from the moon's molten interior, instead of erupting directly, superheats liquid sulphur dioxide just beneath the surface, converting it into gas which bursts from the vents exactly like the pressurised steam of a terrestrial geyser.

Every year, Io pumps about 10,000 million tonnes of matter into space. As it falls back in the moon's low gravity, it coats the surface with sulphur just like the deposits around a Yellowstone fumarole. This is the origin of the satellite's pizza-like appearance. The lurid colours are simply the 'phases' sulphur exhibits at different temperatures.

Jupiter, a whopping 318 times as massive as the Earth, is obviously crucial to understanding Io's super-volcanoes. Io orbits about as far from the giant planet as the Moon is from Earth. But the giant planet's enormous gravity whirls the moon around not in 27 days like the Earth's satellite but in only 1.7 days. That gravity, acting on the tidal bulges of Io, long ago braked Io's rotation so that it orbits today with one face forever locked to its master. If, one day, people get to set foot on that face, what a view they will have, peering through the visors of their spacesuits at Jupiter and its whirling, multicoloured cloud belts filling almost a quarter of the sky.

Because Io's rotation is 'locked', the two bulges pulled in the moon by Jupiter point perpetually towards Jupiter and perpetually away from Jupiter. This means they do not travel through the rock of Io in the way that the tidal bulges on Earth move through the oceans. If such a process occurred on Io, it would stretch and squeeze the rock, over and over again, heating it by internal friction in much the same way that a rubber ball squeezed repeatedly is heated. Since such a process is not happening, it would appear there can be no tidal heating of Io by Jupiter.

But there is.

Critical to Io's heating are two of the other 'Galilean' moons that orbit further out from Jupiter than Io – Europa and Ganymede. Ganymede is actually the largest moon in the Solar System, bigger even than the innermost planet Mercury. For every four circuits Io makes of Jupiter, Europa completes two and Ganymede one. Because of this, the two satellites line up periodically, reinforcing each other's tug on Io. The effect is to yank Io, elongating its orbit. So Io swings in close to Jupiter and then flies back out again, repeatedly. And it is this, it turns out, that is behind the prodigious heating of Io.

Yes, the tidal bulges of Io perpetually point towards Jupiter and away from Jupiter. But, when Io is at its closest to the giant planet, the tidal bulge is bigger than when Io is at its furthest from Jupiter. Up and down, up and down, the rock is stretched and squeezed. The process is so effective at warming the moon that the body in the Solar System currently generating most heat, pound for pound, is not the Sun.[32] It is Io.

The mystery of Pluto and Charon

The Jupiter–Io system is not the only one in the Solar System in which a tidal interaction has resulted in two bodies orbiting each other with their rotations locked so that they perpetually show the same face to each other. There is also Pluto and its giant moon Charon.

The most notable thing about Charon is that it has half the diameter of Pluto. For a while this made Pluto the planet in the Solar System with the biggest satellite relative to its size. But in 2006 it was stripped of its planetary status by the International Astronomical Union and demoted to the category of dwarf planet. It had been found to be merely one of the largest bodies in a swathe of tens of thousands of pieces of icy debris orbiting the Sun in the outer reaches of the Solar System.

The 'Kuiper Belt' is composed of icy builder's rubble left over from the birth of the planets, which could never form a planet because it was spread too thinly. It is the outer Solar System's analogue of the inner Solar System's Asteroid Belt – rocky builder's rubble left over from the formation of the planets, which was prevented from aggregating into a proper world by the gravity of Jupiter. The inner edge of the Kuiper Belt is near Neptune – about thirty times as far from the Sun as the Earth – whereas the outer edge is at about fifty times the Sun–Earth distance. Despite its name, the Kuiper Belt was actually predicted by a former Irish soldier and amateur astronomer called Kenneth Edgeworth in 1943, and, by rights, should be called the Edgeworth–Kuiper Belt.

Pluto fulfils the first two criteria of the IAU's 2006 definition of a planet: it orbits the Sun and is round. But, because it is accompanied by a large number of Kuiper Belt Objects, Pluto does not meet the IAU's third criterion that it should also have cleared its orbit of all other bodies.

On 14 July 2015, NASA's New Horizons sped like an express train through the Pluto–Charon system, skimming just 14,000 kilometres above what had been a planet when the space-probe had been launched a decade earlier but was now a mere dwarf planet. The shock to those back at Mission Control on Earth was that a world they had fully expected to be dead and inert, suspended in the deep-freeze of the outer Solar System, was in fact alive with nitrogen glaciers and mountains of water-ice pushing up towards the thin swirling clouds. Most surprisingly, the pink, heart-shaped region christened 'Tombaugh Regio' after the discoverer of Pluto, Clyde Tombaugh, was devoid of any craters, indicating that ice had spilled across it relatively recently, erasing any sign of the craters that peppered the rest of Pluto.

Where does the energy to drive all this unexpected activity come from? The interior of the Earth is heated by the radioactivity of uranium, thorium and potassium but such heating is believed to be insufficient to warm the interior of Pluto. And tidal heating by Charon is also ruled out since it is impossible in a system in which a moon is in a circular orbit and its rotation is locked to its parent planet. But tidal heating is ruled out only if the capture of Charon took place at the birth of the Solar System, much like the capture of the Moon by the Earth. If, instead, Charon was captured relatively recently – within the past half a billion years – tidal heating could have occurred as the system gradually approached the locked state we see it in today. Nobody knows whether this happened. The jury remains out.

Ocean moons

Tidal heating also has implications for the prospects of life – not on Io perhaps, which seems too inhospitable, but on Europa. Europa is tidally heated by the tug of Jupiter, Io and Ganymede. But, instead of being made of rock like Io, Europa is predominantly made of ice. The unavoidable conclusion is that the interior of Europa must have melted. It must contain liquid water.

A body containing liquid spins differently from a solid body. And the evidence from the way Europa spins is that, beneath a surface layer of ice 10 kilometres thick, there lies a 100-kilometre-deep ocean – the biggest ocean in the Solar System.

From afar, Europa looks like a snooker cue ball, its super-smooth surface the largest ice-rink in the Solar System. But from close-up, giant cracks in the ice come into view. The crazy paving-like pattern of the surface is reminiscent of sea ice in the Arctic Ocean, which breaks up in summer, floats about, then re-freezes in winter. This is yet further evidence of the existence of a sub-surface ocean.

A buried ocean, languishing in permanent sunless darkness, might not seem a likely place to find life. But a key discovery made back on Earth in 1977 has changed everything. Using the submersible 'Alvin', the American oceanographer Bob Ballard discovered 'hydrothermal vents'. Kilometres down on the sea floor, they gush superheated minerals into the ocean and support a thriving ecosystem, all in total darkness. At the bottom of the food chain are bacteria which get their energy not from oxygen but from sulphur compounds. At the top are giant 'tube worms' the size of a forearm.

Given that Europa is tidally heated, there will almost certainly be hydrothermal vents on its sea floor. It makes Europa the most likely place to find life in our Solar System. Currently, NASA is planning to send a space probe to the moon. Ideally, of course, a lander should be dropped onto Europa capable of drilling down through 10 kilometres of ice to the ocean. But this is way beyond current technical capabilities. Nevertheless, Jupiter Icy Moon Explorer (JUICE), planned for launch in 2022, may be able to exploit a recent discovery.

In 2013, the Hubble Space Telescope detected jets of water spouting 200 kilometres into space from cracks in the Europan ice. They can be coming only from a sub-surface ocean. NASA scientists believe that if they fly JUICE through these icy plumes and sample them, they may be able to detect Europan microorganisms.

Another moon with fountains spewing ice into space is Saturn's moon, Enceladus. At barely 500 kilometres across, nobody expected such a tiny moon to be active. But tidal stretching appears to have liquefied its interior. Enceladus may contain the smallest ocean in the Solar System. And, like the ocean of Europa, it may also contain life.

That the Jovian and Saturnian moons are heated by tidal interaction with their parent planets may also have implications for finding life elsewhere in our Galaxy. The reason is that Jupiter and Saturn are outside the 'Habitable Zone' of the Sun. A planet in the Habitable Zone of its parent star is close enough to the star that water does not freeze and far enough away that water does not boil. Jupiter and Saturn are so far away from the Sun that water, essential for 'life as we know it', should freeze. But, as can be seen in the case of Europa and Enceladus, this has not happened. Gas giant planets, many of which are larger than Jupiter, appear to be common around nearby stars. They might very well be orbited by moons bigger than Io and Europa and kept warm by tidal heating.

Precession of the equinoxes

The tides are not the only way that gravity affects the Earth, because it is an extended rather than a point-like object. There is another way discovered by Newton: the precession of the equinoxes.

The seasons occur because the spin axis of the Earth is tilted relative to the plane of its orbit around the Sun. Specifically, as already mentioned, the axis is tilted at 23.5 degrees to the vertical, which means the equator is also tilted at 23.5 degrees to the plane of the Earth's orbit. Summer in the northern hemisphere occurs when the northern hemisphere is tipped towards the Sun and winter when it is tipped away. A similar thing happens in the southern hemisphere. Consequently, it is summer in the southern hemisphere when it is winter in the northern hemisphere, and vice versa.

Spring and autumn are of course the in-between seasons. But astronomers like to be a bit more technical than that. They say that spring and autumn occur when the plane of the Earth's orbit – known as the 'ecliptic' – crosses the plane of the equator. These times during the Earth's journey around the Sun are known as the spring and autumn equinoxes.

All the planets orbit close to the ecliptic because they formed out of a common pancake-like disc of rubble circling the Sun. Consequently, they are confined to a narrow band around the night sky, something that was recognised even in antiquity. In fact, the stars in the band were grouped into twelve constellations, corresponding to the twelve 'signs of the Zodiac'. In 2000 BC, when the Babylonians created the system, the spring equinox was in the direction of Aries. But the spring equinox moves through the signs of the Zodiac at about one sign every 2,000 years. At the time of Christ, the spring equinox was in the direction of Pisces. Today, the spring equinox is beginning to move into Aquarius – it will officially get there in AD 2600 – which is why people talk of the 'coming of the Age of Aquarius'.

This peculiar movement of the Zodiac constellations around the night sky is known as the precession of the equinoxes. It is the third of the Earth's motions – after the planet's rotation on its axis and the planet's circling of the Sun – and it is the most mysterious. Its discovery is credited to Hipparchus, a Greek who lived and worked on the Mediterranean island of Rhodes and who is often described as 'the greatest astronomer of antiquity'.

In 129 BC, when Hipparchus was compiling the star catalogue which made him famous, he noticed an odd thing: the positions of the stars did not match up with the Babylonian measurements. Furthermore, the stars seemed to have shifted their positions in a systematic way. Hipparchus was led to speculate that it was not the stars themselves that had moved but the Earth.

Using Babylonian observations, Hipparchus accurately estimated the speed at which the positions of stars shifted. It seemed that the axis of the Earth changed its orientation in space by about 1 degree every 72 years. This 'precession' causes the planet's spin axis – still maintained at 23.5 degrees to the vertical – to rotate about the vertical once every 26,000 years. Because of this motion, the star immediately above the north pole of the Earth's spin axis – the pole star – is not the same one that the Egyptians saw. We see Polaris in the constellation of Ursa Minor, the Little Bear. But 5,000 years ago the Egyptians saw Thuban, a relatively faint star in the constellation Draco the Dragon.

The wobble, or precession, of the Earth's spin axis is behind the 'precession of the equinoxes'. No one had ever guessed at a reason for this motion. Until Newton.

Newton realised that the shape of the Earth is distorted not only by the gravitational pull of the Moon and Sun but by its own rotation. This causes material on the equator to fly around at 1,670 kilometres an hour. The Earth's gravity has difficulty providing the centripetal force necessary to keep stuff moving around at such a tremendous speed. Consequently, it flies outwards. In fact, the Earth bulges outwards by about 23 kilometres more than it would if the planet were a perfect sphere.

Newton realised that the gravitational pull of the Sun and the Moon on this 'equatorial bulge' causes the spinning Earth to wobble like a top. Indeed, the direction in which the spin axis points moves in a circle. From the forces acting on the Earth's equatorial bulge, Newton calculated that this precession would take 26,000 years, precisely as observed.

Newton's law of universal gravity, it turns out, is a gift that keeps on giving. It has a multitude of consequences. And there are yet more.

In proving the inverse-square law of gravity, Newton assumed that the Sun pulls on each planet as if the Sun and planets are point-like masses. He assumed the Earth pulls on the Moon as if the Earth and Moon are point-like masses. But the Earth and Moon are extended bodies, and this is at the root of other phenomena such as the tides and precession of the equinoxes.

But Newton made another approximation to crank out predictions from his theory of gravity. He assumed that the Sun alone pulls on the Earth, and the Earth alone pulls on the Moon. The breakdown of this assumption can be seen in the case of the tides, where both the Moon and the Sun affect the Earth. And this is a general feature of the real world: bodies are pulled on by more than one other body. This not only leads to bodies such as planets not moving in exact elliptical orbits but to the possibility of deducing from the perturbed motion of known bodies the existence of new ones.

Further reading

'The Severn Bore: A natural wonder of the world' (http://www.severn-bore.co.uk/).

Ekman, Martin, 'A concise history of the theories of tides, precession-nutation and polar motion (from antiquity to 1950)', 1993 (http://www.afhalifax.ca/magazine/wp content/sciences/vignettes/supernova/nature/marees/histoiremarees.pdf).

Shu, Frank, *The Physical Universe*, University Science Books, Mill Valley, 1982.

Simanek, Donald, 'Tidal Misconceptions', 2003 (https://www.lhup.edu/~dsimanek/scenario/tides.htm).

4

Map of the invisible world

*How Newton's law of gravity not only explains
what we see but also reveals what we cannot see*

Kepler's laws, although not rigidly true, are sufficiently near
to the truth to have led to the discovery of the law of attraction of the bodies of the Solar System. The deviation
from complete accuracy is due to the facts, that the planets
are not of unappreciable mass, that, in consequence, they
disturb each other's orbits about the Sun.

Isaac Newton[1]

Pick a flower on Earth and you move the farthest star.

Paul Dirac[2]

They had been searching for almost an hour now and had already slipped into a trance-like automatic rhythm. Johann Galle
peered through the giant brass refractor at the night sky above
Berlin, adjusted the fine control knobs of the telescope until a
star appeared in the crosshairs, and barked out its coordinates.
His younger assistant, Heinrich d'Arrest, seated at a wooden
table across the stone floor of the dome, ran his finger over a
star map by the light of a shielded oil lamp, and shouted back,
'Known star'. Galle twiddled the brass knobs again, lining up
another star. Then another. In the chilly night air, he was already
getting a crick in his neck and beginning to wonder whether they
might have embarked on a futile search.

They were at the Royal Observatory, Berlin, because of an
extraordinary letter that had arrived that afternoon. It was

signed by Urbain Le Verrier, a mathematical astronomer at the École Polytechnique in Paris. Galle had sent Le Verrier a copy of his thesis the year before but had received no reply. Le Verrier, evidently regretting this omission now he needed a favour, had peppered his letter with belated and pointed praise of the Prussian astronomer's work.

It would have been easy for Galle to exact his revenge by accidentally losing the letter among the crowded papers on his desk. He would be doing no more than the French astronomers at the Paris Observatory, who, evidently, had also opted to ignore Le Verrier – otherwise why would he have had to write to Berlin? But Galle was a bigger man than that, and, anyhow, the favour Le Verrier was requesting had piqued his interest. Would Galle use the Observatory's famous Fraunhofer telescope to scan the night sky in the area between the constellations of Capricorn, the goat, and Aquarius, the water carrier? Would he look for an object that did not appear on the star maps?

Joseph Franz Encke, the Observatory's director, thought such a search a waste of time. But that night he would be celebrating his fifty-fifth birthday and so not using the 22-centimetre refractor. Against his better judgment, therefore, and because he could see no possible harm in it, he gave Galle permission to pursue Le Verrier's ridiculous search. Galle quickly roped in an astronomy student, d'Arrest, and here they were, the pair of them, on the morning of Thursday 24 September 1846, scanning the skies with the great clock-driven Fraunhofer telescope, one of the most advanced instruments of its kind in the world.

They had started the search at midnight when the gaslights of Berlin sputtered off, plunging the city into primal blackness, and now it was coming up to one o'clock in the morning. Galle manoeuvred the crosshairs to the next star and called out its coordinates. His mind wandered to thoughts of the warm bed he would soon be in with his wife as he waited for a response from d'Arrest. And waited. What, he wondered, was his assistant doing?

The crash of a chair hitting the floor shocked Galle back to reality. Leaping back from the eyepiece, startled, he saw his assistant, silhouetted against the oil lamp, rushing towards him, flapping his star map like a demented bird. It was too dark to make out the expression on d'Arrest's face but Galle would remember his words for the rest of his life: 'The star is not on the map! It's not on the map!'

Struggling to stay calm and, most importantly, to stop their hands from shaking, the two men took turns at the eyepiece until there could be absolutely no doubt. The object they had found was definitely not on the star map. And the reason for that was clear: it was not a star. Stars, because of their immense distance from the Earth, appear as mere pinpricks of light no matter how powerful a telescope's magnification. But this object was no dimensionless pinprick. It was a tiny shimmering disc.

It was a planet. An unknown planet. Since the birth of the Earth it had been making its way around the Sun in the frigid darkness out at the periphery of the Solar System. And, until now, no one had known. For the moment, it had no name and they were the only two people in the world who knew of its existence. But soon everyone would know it as Neptune.

Pulled in every possible direction

That Galle and d'Arrest should have found the planet was impossible, a piece of magic. A man in Paris had written asking Galle to look for a new world. The man had been very precise about where to look. Intrigued, but not seriously believing he would find anything, Galle had followed the man's instructions. And within an hour, incredibly, the planet had appeared, hovering slap-bang in the middle of the field of view of their telescope, exactly where the man had said it would be. It was a triumph for astronomy. It was a triumph for predictive science.[3] But, most of all, it was a triumph for Isaac Newton and the theory he had devised almost two centuries earlier.

In order to predict things with his universal law of gravity, Newton had made approximations. As mentioned already, he assumed that the Earth pulls on the Moon as if all its mass is concentrated at its centre. In reality, of course, the Earth is an extended body, and differences in the pull of the Moon across its bulk distort its shape, thereby creating the phenomenon of the tides. But the assumption that the Earth pulls on the Moon as if its mass is concentrated at its centre is not the only approximation made by Newton. He also assumed that a planet experiences a gravitational pull from only the Sun. By making this assumption, he was able to prove that the path of a planet moving under the influence of a force that weakens with distance according to an inverse-square law is an ellipse, exactly as Kepler had found.

But the central characteristic of gravity is that it is a universal force, which means every mote of matter in existence exerts a gravitational pull on every other mote of matter. This means that a planet as it travels through space is tugged not only by the Sun but also by every other planet. Take the Earth. The two other planets which exert the greatest pull on our planet are Jupiter, the biggest world in the Solar System, which weighs in at about 1/1,000th the mass of the Sun, and Venus, the closest world to our own. The pull of each on the Earth varies with time because Jupiter travels around the Sun more slowly and Venus more quickly than the Earth. But when Jupiter is at its closest, it exerts a pull which is about 1/16,000th that of the Sun. And when Venus is at its nearest, it exerts a pull about half of that.

That each planet experiences much smaller gravitational tugs from the other planets in the Solar System than it does from the Sun was the reason Newton was able to ignore them in his calculations. But, strictly speaking, a planet like the Earth moves under the influence of multiple bodies. And a consequence of this is that, actually, it does not orbit the Sun in a perfect ellipse. Kepler's first law is only approximately true. The effect of all the other tugs is to cause the elliptical orbit of a planet gradually to change its orientation in space over a long period of time, with the closest point in the orbit to the Sun creeping, or 'precessing', around the Sun.

Say for some reason we are totally ignorant of the existence of the other planets of the Solar System. After monitoring the Earth's orbit for a long time, we discover that it is deviating ever so slightly from a perfect ellipse. After considering the matter, we conclude that there are other massive bodies out in space tugging at the Earth as it flies along rather like children tugging at their mother's coat as she hurries along a road. Employing enough computing power, not to mention a lot of ingenuity – the calculations are not easy – we deduce that the Earth is being tugged by seven other planets, each with a very particular mass and orbiting at a very particular distance from the Sun.[4]

Newton's law of gravity has enabled us to make a map of the invisible world of our Solar System. And this technique is precisely the one Le Verrier used to map the outer reaches of known space and deduce the location of an undiscovered eighth planet, Neptune. This was because there was a planet whose orbit was deviating from a perfect ellipse.

A planet called George

Uranus had been discovered by William Herschel, a freelance German musician. In 1757, at the age of nineteen, he and his sister Caroline moved to Bath, a town in England founded by the Romans because of its hot springs.[5] Herschel found work as a church organist. But his passion was for astronomy. And in the garden of his house, he built one of the best telescopes of his day. It was while scanning the night sky with this instrument, on 13 March 1781, that a fuzzy star popped into his eyepiece. At first Herschel thought it was a comet. But, over successive nights, as it crept across the constellation of Gemini, Herschel realised it was not following the highly elongated orbit of a comet but the near-circular orbit of a planet.

The new planet caused an international sensation. From the beginning of recorded history people had known of only six planets. Now there were seven.

As an immigrant, Herschel's greatest desire was to be accepted in his adopted country. So he christened the new planet 'George' after King George III (actually, he named it 'George's star'). This upset the French astronomers, who strongly objected to a planet named after an English king and instead referred to it as Herschel.

In a rare instance of the Germans acting as peace-makers, the astronomer Johann Bode suggested Uranus, the father of the Roman god Saturn. The name stuck. If it had not, then, moving outwards from the Sun, we would today have Mercury, Venus, Earth, Mars, Jupiter, Saturn . . . and *George*.

Uranus had actually been seen almost a century earlier by English astronomer John Flamsteed. In 1690, mistakenly believing it to be a star, he had catalogued it as 34 Tauri, the thirty-fourth star in the constellation of Taurus. The historic records of Flamsteed and others of Uranus's position, supplemented by new observations, meant that, by the early nineteenth century, the orbit of Uranus was known precisely enough that it could be compared with the prediction of Newton's law of gravity.

But this threw up a puzzle.

The same elliptical orbit did not fit all of the observations. As soon as an orbit was determined for Uranus, the planet began wandering from its path. And as the decades passed and more observations were made, Uranus strayed more and more.

Few doubted that there could be anything wrong with Newton's law of gravity. Its successes over the past couple of centuries had been so overwhelming and so comprehensive that it had achieved status akin to the word of God. Instead, the suspicion arose that there must be another world beyond Uranus whose gravity was tugging at the planet and causing it to veer from a neat elliptical orbit.

Hunting the invisible planet

In 1841, John Couch Adams, an autistic mathematical genius from Cornwall in England, set out to deduce where in the sky the new planet must be in order to have the observed effect on Uranus.[6] His calculation was horrendously complicated. But, four years later, in 1845, he was able to take his result to the Astronomer Royal, George Challis. Challis did not take him seriously. It did not help Adams' credibility that he kept refining his calculations and presenting Challis with new and slightly different predictions of where to look for the new planet.

Meanwhile in France, and unknown to Adams, Le Verrier was carrying out similar computations. In order to simplify the intimidating calculations, he made a number of educated guesses. He assumed, for instance, that the unknown planet was a long way from the Sun or else it would have been spotted already by astronomers; he assumed it had a comparable mass to Uranus, which is about fifteen times as massive as the Earth; and he assumed that it orbited the Sun in the same plane as the other planets.[7]

Coincidentally, Le Verrier faced similar obstacles to Adams in being taken seriously. The director of the Paris Observatory, Françoise Arago, did not see a search for a new planet as at all urgent. Unable to pin down Arago on a date to start looking, Le Verrier lost patience, and, on 18 September 1846, sent his estimate of the rough location of the new planet to Berlin. Five days later, the only man to trust Le Verrier, Johann Galle, entered the history books as the discoverer of Neptune.

Like Uranus, Neptune had been seen before but not recognised as a planet. It was actually just about visible with the naked eye. In fact, there is some evidence that, as early as December 1612, Galileo, in Padua, had seen it through his new-fangled telescope, mistaking it for a star.

The discovery of Neptune triggered a priority dispute between France and England. Remarkably, it did not spill over into the relationship between Adams and Le Verrier, even though Le Verrier had a reputation for arrogant and bullying behaviour. Perhaps because they appreciated each other's mathematical wizardry and because they had faced similar obstacles in getting mere mortals to believe them, they became firm friends as soon as they met. Nowadays, often as not, the discovery of Neptune is attributed jointly to Adams and Le Verrier.

The discovery of Uranus had been a sensation. It was the first planet discovered in the age of the telescope, the first in the age of science. It was twice as far from the Sun as Saturn, so overnight it doubled the size of the Solar System. The discovery of Neptune was a sensation of an entirely different order. Whereas Uranus had been stumbled on by accident, Neptune's existence

– its mass, its appearance, its very location – had been predicted. Science had given men the power of gods. Not only did Newton's law explain what we could see, it also predicted what we could not see.

And in the twenty-first century history may be repeating itself.

Planet Nine

At the start of 2016, two planetary scientists in the US stunned the scientific world by claiming that a hitherto undetected planet about ten times the mass of the Earth is orbiting the Sun far beyond the outermost planet. For want of a better name, Konstantin Batygin and Mike Brown of the California Institute of Technology in Pasadena have christened the new world 'Planet Nine'. Pluto was of course the ninth planet until its ignominious demotion to a dwarf planet in 2006.[8]

The evidence cited by Batygin and Brown is not the anomalous motion of a planet but the anomalous motion of Kuiper Belt Objects. As mentioned earlier, these bodies – icy builder's rubble left over from the birth of the planets – orbit in their tens of thousands just beyond the outermost planet, Neptune.[9] Batygin and Brown note that six of the most distant KBOs have highly elongated orbits which are roughly aligned. Rather than pointing in random directions, as might be expected, they all point more or less in the same direction. The orbits are also tilted in the same way, pointing about 30 degrees downwards from the plane of the eight known planets. The best explanation for the anomalous motion of these KBOs, according to Batygin and Brown, is that all are being herded by the gravity of the distant and unseen planet.[10]

Not only must such a planet be extraordinarily big, it must also be extraordinarily far away: on average about twenty times as far from the Sun as Neptune. Batygin and Brown estimate that Planet Nine follows a highly elongated orbit, swinging in to about seven times Neptune's distance from the Sun before flying out to about thirty times the distance of the outermost planet. Travelling in such a huge orbit, it takes not 165 years to go around the Sun, as does Neptune, but about 15,000 years.

Planet Nine could have formed along with the other planets 4.55 billion years ago and then been catapulted out into the cold by a close encounter with an embryonic giant planet such as Jupiter or Saturn. Or there is a small possibility that it is a planet captured from another star. In the stellar nursery where the Sun was born there would have been hundreds of other suns in close proximity, and encounters between them could conceivably have switched planets. The possibility that there exists an alien planet in our Solar System is a reminder that science is often stranger than science fiction.

At its predicted distance from the Sun, Planet Nine is expected to reflect little sunlight, making it very difficult to find even with the biggest telescopes. If the planet happens to be at its closest point to the Sun in its orbit, it should be possible to spot it in existing celestial images, captured by previous surveys of the night sky. If it is in the most distant part of its orbit, spotting it will require the world's largest telescopes such as the twin 10-metre telescopes at the Keck Observatory on Mauna Kea in Hawaii. But Planet Nine – which is estimated to be about 3.7 times the diameter of the Earth with a chilly surface temperature of −226 degrees Celsius – might be easier to detect with infrared telescopes sensitive to its meagre heat output.

If Planet Nine exists, it may make our Solar System more like the 2,000 or so planetary systems so far found around other stars. One of the most common types of extrasolar planet has a mass between that of the Earth and Neptune, which is seventeen times heavier than the Earth. If such a 'Super Earth' once existed but got kicked out into the cold, it would explain why our Solar System apparently lacks such a world.

Ironically, Brown played a key role in the demotion of what was the ninth planet, Pluto, to its dwarf planet status. It was his 2005 discovery of Eris, a remote icy world nearly the same size as Pluto, that showed that what had been considered the outermost planet since 1930 is merely the biggest of many bodies in the Kuiper Belt. By proposing a possible replacement for Pluto, Brown is making amends for his murder of a planet.

Of course, Planet Nine may turn out to be an illusion. And some astronomers remain sceptical. Nevertheless, the power of Newton's law of gravity to predict what we cannot see – to make a map of the invisible world – has continued to reap dividends.

Exoplanets

Currently, we know of several thousand worlds in orbit around other stars. Hardly any of these 'exoplanets' have been seen directly. Instead, the existence of many of them has been inferred indirectly from the effect they exert on their parent suns. It is all down to the mutuality of Newton's law of gravity. A sun pulls on a planet with exactly the same force that a planet pulls on its sun. Of course, the sun, being hugely more massive and more difficult to budge, moves relatively little in response. Nevertheless, it moves.

What this means is that, strictly speaking, a planet does not orbit a static sun – that's another one of the approximations used by Newton in order to squeeze out predictions. Instead, both a planet and its sun orbit their common centre of mass. Since a sun is a lot more massive than a planet, this centre is close to the centre of the sun, and usually well inside its fiery body[11] While the planet orbits the centre of mass of the system in a big orbit, the sun orbits the centre of mass in a tiny orbit.

Another way to think of it is that a planet tugs on its sun in one direction, then half an orbit later, tugs it in the opposite direction. This causes the sun to move a little, or 'wobble', and this is detectable from Earth with sensitive instruments. In exactly the same way that the frequency, or 'pitch', of a police siren gets higher as it approaches and lower as it recedes, the frequency of light emitted by atoms in a star becomes higher or lower depending on whether a star is approaching or receding from the Earth. By measuring the magnitude of this 'Doppler shift' for common atoms of elements such as hydrogen, astronomers can determine the velocity of a star towards us and away from us.

In the case of a star being tugged by the gravity of a planet, the maximum wobble velocity is only a few metres per second for a planet the mass of Jupiter and a mere 10 centimetres per second if it is the mass of the Earth. In other words, a ball of hot gas, typically more than a million kilometres across, approaches and recedes from us at the speed of a jogger and a tortoise, respectively. Despite the formidable technical challenge, astronomers are able to measure such velocities using a super-sensitive 'spectrograph' and infer the existence of an invisible planet.[12] In this way, more than 2,000 have been discovered since the mid-1990s, and the hunt is on for a second Earth.[13]

The most dramatic example of the power of Newton's law of gravity to predict what we cannot see is not in the realm of stars and planets but in the domain of the large-scale Universe. In the late twentieth century, astronomers discovered, to their astonishment, that the stars and galaxies, which they had believed were the sole cosmic building blocks, are in fact only a minor constituent of the cosmos. There is an awful lot more stuff out there than we ever imagined. And it is entirely invisible.

Invisible Universe

In the late 1960s and 1970s, astronomers Vera Rubin and Kent Ford at the Department of Terrestrial Magnetism of the Carnegie Institution of Washington began studying spiral galaxies. These great whirlpools of stars account for about 15 per cent of all galaxies and of course include our own Milky Way. Rubin and Ford set out to measure how fast the stars in spiral galaxies are whirling around their massive central 'bulges'.

The two astronomers picked spiral galaxies seen edge-on because the stars in such systems are moving along our line of sight. With a super-sensitive spectrograph, they were able to measure the velocity of the stars more accurately than anyone had ever done before.

At greater and greater distances from the centre of each galaxy, the stars should be experiencing ever weaker gravity. Consequently, Rubin and Ford expected to find the stars orbiting ever more slowly, exactly as the planets in our Solar System do at greater and greater distances from the Sun.

But this is not what they found.

As far out from the centre of each spiral galaxy as it was possible for the two astronomers to see stars, the orbital velocity of the stars remained constant. The stars were whirling round far too fast. Like children on a sped-up merry-go-round, they should be flung off their galactic carousel. They should be sailing out across intergalactic space. There was no way that the gravity of their parent galaxies could be holding on to them.

But it was.

Modern-day astronomers, in common with their nineteenth-century predecessors, have an unshakeable faith in Newton's law of gravity, which has scored so many successes over so many centuries.[14] So they have come up with an explanation for the anomalous motion of the stars in spiral galaxies that is not a million miles from Adams and Le Verrier's explanation of the anomalous motion of Uranus. The reason the stars in spirals do not fly off into intergalactic space, maintain astronomers, is that they are in the grip of the gravity of more matter than is visible with telescopes. A lot more.

Remarkably, every spiral galaxy is embedded in a vast spherical cloud of 'dark matter', rather like a CD embedded in a swarm of bees. The dark matter gives out no light – or at least insufficient light to be detectable by the most sensitive instruments currently available – and outweighs the visible stars typically often by a factor of 10.

The discovery of Neptune merely showed that we had overlooked the existence of a planet in the Solar System. The discovery of dark matter is a bit more serious than that. It shows we have overlooked most of the matter in the Universe.

Actually, the first hints that there is more to the Universe than anyone suspected came in the 1930s. Fritz Zwicky, a Swiss-American astronomer at the California Institute of Technology in Pasadena, was observing galaxy clusters. To his surprise, he discovered that their constituent galaxies are being whirled around so fast they should be flung off to infinity. In Holland, at roughly the same time, Dutch astronomer Jan Oort discovered that stars in the vicinity of the Sun appear to be orbiting faster around the centre of our Milky Way than can be explained by the gravity of the visible matter inside the Sun's orbit.

Zwicky concluded that there must be a lot more matter in galaxy clusters, and Oort that there must be a lot more matter in our own Galaxy, than shows up in telescopes. It is the extra gravity of this dark matter – a term actually coined by Zwicky (*Dunkle Materie* in the German) – which must hold onto their galaxies and stars, respectively.

The idea that there is 'missing mass' in the Universe somehow did not enter the mainstream of astronomy, maybe because it was so hard to believe. But everything changed with Rubin and Ford's super-precise observations of stars orbiting in spiral galaxies.[15] Here were anomalous measurements of the velocities of stars – lots of them – which could not be swept under the carpet.

Gravity can do more than reveal the presence of dark matter; it can be used to deduce its distribution as well. This is because the light from distant galaxies, on its journey to Earth, is bent, or 'lensed', by the gravity of the dark matter it passes. From the distortion, or 'weak lensing', of the images of the distant galaxies it is possible to deduce the distribution of that dark matter. Currently under construction on a mountaintop in Chile is a telescope that can exploit this effect. The Large Synoptic Survey Telescope is a kind of anti-telescope that turns the idea of a telescope on its head.[16] By means of the light it collects, it images darkness.

The evidence for the existence of dark matter comes not only from spiral galaxies but also from another important place. The Universe was born 13.82 billion years ago in a titanic explosion – the big bang – and has been expanding and cooling ever since. Out of the cooling debris there congealed 100 billion or so galaxies, including our own Milky Way. One serious problem with this picture is that it fails to predict a rather important feature of the Universe: that we exist.

Galaxies formed because some regions of the big bang fireball were slightly denser than others. (The 'density fluctuations' are believed to have been imprinted on the Universe by 'quantum' processes in the first split-second of creation – but that is another story.)[17] Because the slightly denser regions had slightly stronger gravity, they dragged in material faster than other regions, which boosted their gravity so they pulled in matter yet faster in a process akin to the rich getting ever richer. But the point is this: the process is too slow. The 13.82 billion years which have elapsed since the moment of the Universe's birth is woefully too short a time to have built up galaxies as big as the Milky Way. Unless there exists a lot more matter than we can see with our telescopes – matter whose gravity speeded up galaxy formation. Dark matter.

The Universe's dark matter outweighs the visible matter – the stuff made of atoms like you, me, the stars and galaxies – by a factor of between five and six. In fact, because of a European space telescope called 'Planck', which observed the 'afterglow' of the big bang fireball, we can be even more precise. Whereas 4.9 per cent of the mass-energy of the Universe is atoms, 26.8 per cent is dark matter. (The remaining 68.3 per cent, known as 'dark energy' and discovered only in 1998, is invisible, fills all of space and has repulsive gravity – but that too is another story.)[18]

As for the identity of the dark matter – what it actually *is* – your guess is as good as mine. One idea is it is made of hitherto undiscovered subatomic particles. Speculative theories of physics such as 'supersymmetry' postulate the existence of a host of new fundamental particles which do not 'feel' the electromagnetic force and so generate no electromagnetic waves, also known as light. Another idea is that the dark matter is made of fridge-sized black holes each as massive as Jupiter which were created in the violent conditions that existed in the big bang fireball.[19]

If the dark matter is made of 'primordial' black holes, and assuming they are spread uniformly throughout the Universe, the nearest would be about 30 light years away, about ten times further away than the nearest star, Alpha Centauri. If the dark matter is made of subatomic particles, then they are flying through you at this very instant, as unaffected by the atoms of your body as bullets flying through fog. Only one thing is certain about dark matter: if you can discover its identity, there is a Nobel Prize waiting for you in Stockholm.

Using modern parlance we can now say that Neptune was the

dark matter of its day. But, if we took a time machine back to the nineteenth century, we would discover that it was not the only dark matter planet. There was another, rather ghostly and slippery one. Its name was Vulcan.

Vulcan

You may be forgiven for thinking Vulcan is the ancestral home of the ultra-logical Mr Spock from *Star Trek*. But Gene Roddenberry, the creator of the 1960s American TV series, did not conjure the name from thin air. The planet already existed. Or at least it existed in the imagination of nineteenth-century astronomers, most notably Le Verrier.

After his triumphant prediction of the existence of Neptune, Le Verrier's star rose in the scientific firmament, and, in 1854, he became director of the Paris Observatory. But nothing he did, nothing he achieved, came close to matching the sheer blood-pumping exhilaration he had felt at magically unveiling an unknown world at the edge of the Solar System. For his achievement, he had been courted by kings, revered as a god by scientists. Fame and adulation had intoxicated him and he craved that feeling again. If only he could repeat his success. If only he could make another god-like pronouncement that would stun the human world. And so his attention turned from the outer to the inner Solar System.

Le Verrier's goal was characteristically ambitious: to completely understand the orbits of the inner planets: Mercury, Venus, Earth and Mars. If he could do that, then, perhaps, just perhaps, an anomaly would show up that would lead to a new, headline-grabbing discovery.

Each planet, as pointed out before, is influenced not only by the gravity of the Sun but by the gravity of all the other planets. As a result, it does not actually trace and retrace the same path for all eternity. Instead, over long periods of time, its elliptical orbit precesses, causing the planet to trace out a rosette-like pattern in space. Because precession causes the closest approach of a planet to the Sun, known as its 'perihelion', to gradually circle the Sun, astronomers talk of the 'precession of the perihelion' of a planet.[20]

It was in 1843, three years before the discovery of Neptune, that Le Verrier had first focused his attention on the four innermost planets. In order to predict the orbit of each world, he had painstakingly added up the gravitational tugs from all the other planets in the Solar System. Unfortunately, his predicted orbits did not match the observed orbits. He suspected that the discrepancies were due to his imperfect knowledge of the distances and masses of the other planets. And so, in the decade after his Neptune triumph, he set himself the task of refining those planetary vital statistics.

In 1852, the best estimate of the average distance between the Earth and the Sun was 95 million miles. By 1858, Le Verrier had refined the figure to 92.5 million miles, which is within half a per cent of the modern value. The following year, armed with this much-improved figure, Le Verrier set out once more to calculate the orbits of the inner planets.

It was a long and tedious marathon of number crunching. And Le Verrier had no more success than he had had sixteen years earlier. The orbits he calculated for the innermost planets did not match the orbits observed by astronomers. But he had faith in Newton's law of gravity, he believed in his mathematical intuition, and so he persevered with his calculations. It seemed likely to him that the problem was with the numbers he was using for the masses and distances of the planets. Perhaps they were still in error. He tried adjusting them, one at a time. It took a while to do this. But, eventually, his efforts paid off. All that was needed was a simple change. After slightly increasing the masses of the Earth and Mars, he was able to predict the precise orbits of all of the inner planets.

All the planets, that was – except one.

Mercury is the innermost planet, orbiting closest to the fires of the Sun. It is also the tiniest planet, smaller even than Jupiter's moon, Ganymede.

According to Le Verrier's calculations, the pull of Mercury's closest planetary neighbour, Venus, causes the planet's perihelion to advance by about 1/5,000th of the way around the Sun every century. Astronomers use an even more esoteric and opaque language than this. They say Venus causes Mercury's perihelion to advance by 280.6 arc seconds per century, with an arc second being 1/60th of an arc minute, and an arc minute 1/60th of a degree. Le Verrier's calculations showed that the tug of the giant planet, Jupiter, contributes another 152.6 arc seconds per century; the Earth 83.6 arc seconds per century; and all the remaining planets combined a mere 9.9 arc seconds per century. Adding together all these numbers, Le Verrier arrived at a figure for the precession of the perihelion of Mercury of 526.7 arc seconds per century.

But this was not right.

Careful observations of the inner planet had shown that Mercury's perihelion advances by 565 arc seconds per century. This left a discrepancy of about 38 arc seconds per century (the modern value is 43 arc seconds per century).

The mismatch was tiny. But Le Verrier's calculations were precise enough to show that it was real. Mercury's perihelion was precessing by 38 arc seconds per century more than it should. In other words, if all the other planets in the Solar System were to vanish suddenly, removing at one fell swoop their long-range gravitational effects, Mercury would still trace out a rosette pattern. A rosette that repeats roughly every 3 million years. A rosette pattern that is utterly inexplicable.

Le Verrier could hardly believe his luck. It was the Uranus anomaly all over again. A hidden mass – something within the orbit of the inner planet – was tugging on Mercury. Le Verrier hardly dared voice the question. But could it be, was it possible that it was, a new planet?

To estimate its mass, Le Verrier assumed that it orbited halfway between Mercury and the Sun. His calculations showed that such a planet could account for Mercury's anomalous precession if its mass was similar to its neighbour. But that immediately posed a problem. A planet that big should long ago have been spotted by astronomers. Yes, it would be hidden in the glare of the Sun. But it should have shown up during a total eclipse when the Moon blots out the Sun and even faint stars can be seen close to the solar disc.

If it was not a planet, what might it be? Le Verrier wondered whether Mercury's wayward behaviour might instead be caused by a collection of 'asteroids' orbiting between the Sun and Mercury. If this was the case, then some of the objects might be big enough to be seen as they crossed, or 'transited', the face of the Sun.

Incredibly, someone already had spotted something transiting the Sun. Edmond Modeste Lescarbault was a French country doctor with a passion for astronomy. He had been thinking about the asteroids which had been discovered orbiting between Mars and Jupiter in the early decades of the nineteenth century, and this had got him wondering where else asteroids might lurk.[21] With his 4-inch refractor telescope at Orgères-en-Beauce, about 70 miles west of Paris, he had previously observed the tiny black speck of Mercury transiting the Sun. So it was natural for him to wonder whether there might exist asteroids closer to the Sun than Mercury, and whether he might be able to spot them as they crossed the solar disc.

On Saturday 26 March 1858, Lescarbault was running his surgery. But there was a gap between patients. So he took the opportunity to go to his telescope and point it at the Sun. To avoid blinding himself, he projected the image of the solar disc onto a card. As soon as he did so he saw something unusual: a small black dot close to the edge of the Sun. He was of course desperate to follow its progress but another patient had arrived and was demanding his attention. When he finally dashed back to his telescope, he was relieved to see that the dot was still there. Lescarbault tracked it continuously until it vanished off the other edge of the Sun. The total time it had taken to transit, he estimated, was 1 hour 17 minutes and 9 seconds. It was exactly what would be expected for an asteroid in the innermost reaches of the Solar System.

Curiously, Lescarbault did nothing about his discovery. It was only nine months later, when he read an article saying that Le Verrier believed there was a body or bodies between Mercury and the Sun, that he put pen to paper and wrote to the Paris Observatory.

Le Verrier was deeply sceptical of the doctor's claim. But the possibility that he, Le Verrier, might repeat his Neptune success was simply too tantalising. He had to meet Lescarbault. On 31 December 1859, he caught the train from Paris to Orgères-en-Beauce. Arriving unannounced at Lescarbault's home, he fully expected to find an unimpressive rural amateur. Instead, he met a first-rate observer who had built precision scientific instruments. After grilling him exhaustively on his observations, the Parisian astronomer became convinced of his discovery.

Unbelievably, Le Verrier had done it again. He had repeated his success with Neptune. He had predicted the existence of a planet between Mercury and the Sun. He was truly a god among men.

Back in Paris, Le Verrier converted Lescarbault's observations into numbers. Assuming the new planet was travelling in a circular orbit around the Sun, it should complete a circuit once every twenty days. That meant it should transit across the face of the Sun several times a year as seen from the Earth.

Le Verrier announced the discovery of the new planet to an astonished world. By February 1860 it even had a name. Planets are named after ancient gods, and the lord of the forge on Mount Olympus, the home of the Greek gods, was Vulcan. It seemed an entirely appropriate name since the new world could never escape the fires of the Sun. So Vulcan it became.

Other astronomers, particularly those who monitored the Sun for sunspots, quickly announced that they had also seen Vulcan transiting the Sun but had not recognised it as a planet.[22] There was another opportunity to observe a transit between 29 March and 7 April 1860. In Madras in India, and in the Australian cities of Sydney and Melbourne, astronomers watched the Sun's disc continuously. But nothing showed up.

The years passed. Some observers claimed to see the new planet. Many others did not. And the observations of those who saw something never seemed to be independently verified by anyone else.

On 7 August 1869 there was a total eclipse. Once again, some observers reported that they saw Vulcan. But, crucially, the total eclipse was observed by an American pioneer of astrophotography from Burlington, Iowa. Benjamin Apthorp Gould took forty-two photographs of the misty white 'corona' that surrounds the Sun, and which is visible only during totality. None showed the new planet.

The clincher was the total eclipse of 29 July 1878. Teams of astronomers took the Union Pacific railway to Rawlings, Wyoming, in the American Midwest. Among them were some of the greatest observers of the day. They included Simon Newcomb, from the Naval Observatory in Washington, DC, who was destined, unfortunately, to be remembered for declaring heavier-than-air flight impossible on the eve of the Wright brothers' pioneering flight; and Norman Lockyer, who, from his garden in the London suburb of Wimbledon on 20 October 1868, had discovered helium on the Sun, the only element to be discovered in space before it was discovered on Earth. Even the world-famous inventor Thomas Edison tagged along.

From Rawlings, the observers tramped with their equipment to suitable observing spots. They were plagued by cloudy skies and eye-watering dust and sand, whipped up by the incessant wind. But, despite all the meteorological difficulties, not to mention malfunctioning equipment, many saw and even photographed the eclipse. Only one man saw a new planet.

James Craig Watson, director of Michigan's Ann Arbor Observatory, reported a small ruddy object, orbiting the Sun inside the orbit of Mercury. His discovery was immediately wired around the world. Two decades after Le Verrier had proposed the existence of a new planet, could it be that Vulcan had at last put in an appearance?

The trouble was nobody else saw it. Or, rather, they saw the ruddy speck but recognised it as Theta Cancri, a faint star in the constellation of Cancer the Crab. Watson stuck to his guns even when it seemed overwhelmingly likely he was wrong and everyone else was right. In fact, when he went to his grave in 1880, having contracted a fatal infection at only forty-two, he was still utterly convinced he had discovered the planet Vulcan.

But the balance had now tipped. The consensus was that Vulcan did not exist, had never existed. It was a figment of the fevered imagination. A testament to the power of human delusion, of scientific wishful thinking. It lived on only as a half-forgotten historical footnote and, of course, the fictional birthplace of *Star Trek*'s Mr Spock.

Unsolved puzzle

The idea of a planet like Vulcan turns out to be not so mad after all. Thousands of planets are now known to orbit other stars in our Milky Way, and many of them have Vulcan-like planets.

One of the most unexpected discoveries in modern astronomy is of gas giant planets orbiting their parent stars closer in than Mercury orbits the Sun. Such 'hot jupiters' could not possibly have formed where we see them. The gas would have been so hot, its constituent gas atoms flying about so fast, that gravity could not have held onto it. Instead, astronomers believe hot jupiters are born much further out. Friction with the debris disc out of which planets form causes them to spiral inwards. Such planetary 'migration' is now believed to have also been a feature of our Solar System's pre-history, with worlds like Jupiter and Saturn having participated in a game of interplanetary musical chairs before taking up their current locations.

Planetary systems around other stars appear to be telling us that our Solar System is unusually stretched out. More than half the planets in 'exoplanetary' systems orbit closer to their parent star than Mercury is to the Sun. Elsewhere in our galaxy, Vulcans abound. It is still possible that this is an illusion caused by 'observational biases'. Astronomers detect exoplanets by the wobble they create in their parent star or by the dimming they produce in their parent star's light. Close-in planets are quicker and easier for astronomers to spot since there is less time to wait for them to complete an orbit.

Our planetary system may not always have been so unusual. According to computer simulations of the birth of the Solar

System, initially there may have been a number of worlds orbiting close in to the Sun. Collision between them left Mercury as the sole survivor. If this scenario is correct, then Vulcan did indeed exist. Unfortunately, the human race missed it by 4.55 billion years.

Le Verrier died on 23 September 1877. He had solved the problem of the anomalous motion of Uranus, discovering Neptune and in the process expanding the size of the Solar System. But, with Vulcan slipping inexorably from his grasp, he knew that the problem of the anomalous motion of Mercury had defeated him.

The twentieth century arrived and there were marvels galore to attract attention: X-rays and radioactivity and human-powered flight. The anomalous motion of Mercury was a curious puzzle but it was almost certainly not an important one. Nobody lost much sleep worrying about it. In fact, nobody gave it much thought at all. And nobody suspected what it was really telling us: that, incredibly, impossibly, Newton was wrong about gravity.

The man who recognised this and devised a better theory of gravity to supplant Newton's was Albert Einstein. But, before it dawned on Einstein that his predecessor had been wrong about gravity, he realised that Newton was wrong about something apparently even more fundamental that had a bearing on gravity: the very nature of space and time.

Further reading

Aw, Tash, *Map of the Invisible World*, Fourth Estate, London, 2010.

Levenson, Thomas, *The Hunt for Vulcan ... And how Albert Einstein destroyed a planet, discovered relativity and deciphered the Universe*, Head of Zeus, London, 2015.

Schilling, Govert, *The Hunt for Planet X*, Copernicus Books, New York, 2009.

PART TWO

Einstein

Catch me if you can

*How Einstein realised that nothing can travel
faster than light and that this is incompatible
with Newton's law of gravity*

For Mr Newton, space and time did not talk to each other,
never married, and lived separate lives.

Roberto Trotta[1]

The velocity of light in our theory plays the part, physically,
of an infinitely great velocity.

Albert Einstein[2]

'What would it be like to catch up a light beam?' Einstein was
just sixteen when he asked the question that would set him on the
road to greatness. Frustratingly, he never told anyone the exact
circumstances which prompted him to ask this critical question.
Instead we can only speculate. We know that he formulated
the question in early 1896 while at school in the Swiss town of
Aarau, 30 miles west of Zurich. At the time, he was boarding
with the Winteler family.

I imagine him waking to sunlight streaming through the
window of the attic room he rents. The swaying branches of a
linden tree splinter the light into a myriad glowing fragments,
which dance kaleidoscopically on the wall beside his bed. He
reaches up with his hand and, like a child, tries to grab a jittering
lozenge of light. So utterly transfixed is he by the shifting shapes
on the wallpaper that he makes no move to push back his covers
until the spell is broken by a rap on the door. 'Herr Einstein!' It

is Marie Winteler, the pretty eighteen-year-old daughter of his landlord, who has taken a fancy to him. 'Papa says to tell you breakfast is ready.'

I picture Einstein, later that day, sitting at a desk in a high-ceilinged classroom at the Aargau Cantonal School, staring idly out at the River Aare. The rain, which has been spattering the window, stops as abruptly as it started. As the thick clouds part, the night-time gloom which has descended on the little Swiss town is pierced by a biblical shaft of light, stabbing down as if from Heaven. Where it strikes the black river, the water sparkles like a nest of diamonds. So mesmerised is he by the sight that he hears nothing of the lecture – on wiring patterns of AC dynamos – until his reverie is interrupted by the roar of Doktor August Tuchschmid, the school's principal. 'Herr Einstein! My sincerest apologies for boring you. Perhaps, at some time in the half hour that remains to us, you might deign to honour us with your attention.'

That night I see Einstein and Marie Winteler running hand in hand through the narrow lanes of Aarau, splashing through puddles and laughing hysterically like the teenagers they are. Soaked through but not caring, they stop abruptly and he pulls her towards him and kisses her. Over her shoulder, he sees the eerie-green globes of the gas lamps, marching down the street, growing steadily smaller and fainter as they converge in the distance. In the oily black puddles he sees their reflections, and also the reflection of the full Moon, like some rogue lamp that has broken free of the Earth and floated up to the sky. He stops kissing her and looks upwards.

'Albert?'

All day he has been mesmerised by light. All day he has been wondering about light. All day he has been asking himself the same pressing question: what is wrong with our understanding of light? In the question lies the answer. But his question is too woolly, too imprecise, to make any progress.

His girlfriend has spoken to him but, in his mind, he is a quarter of a million miles away. The light of the Moon has travelled that far across space to reach his eyes. He pictures the journey it takes – barrelling along through the cold vacuum at a billion kilometres an hour – and his heart misses a beat. Suddenly, he knows the question he should be asking – the one that has the potential to open the doors to a new world of understanding. It is so obvious he cannot believe it has never occurred to him before.

'Albert, what are you thinking?'

Even before he answers, she knows it will be something she would never, ever have guessed. Although he is just sixteen, already he sees the world differently from other people, thinks different thoughts from everyone else. The textbooks she has seen him studying in his room until the early hours might as well be written in hieroglyphics. She knows she cannot go where he goes, that she cannot enter his world. She has a sudden premonition that she will soon begin to bore him and that he will leave her, and a tear forms in her eye.

'What am I thinking?' he says, as if waking from a dream.

'Yes.' She wipes her tear on the sleeve of her coat, but he does not notice.

'I was thinking: what would it be like to catch up a light beam?'[3]

Rolling her eyes, she takes his hand and pulls him towards home.

'Albert, you are so strange.'

Of course, this is fantasy. But it is fun to imagine! When the sixteen-year-old Einstein formulated his critical question, light was known to be a wave like a wave rippling over the surface of a pond. This is not an obvious fact because the distance between successive crests of a light wave is very small – much less than the width of a human hair. But an ingenious experiment carried out by an English physician called Thomas Young in 1801 had confirmed the wave nature of light.[4] Still, nobody had the slightest idea what light *was*.

Everything changed in 1863. In a theoretical *tour de force*, the Scottish physicist James Clerk Maxwell succeeded in summarising all electric and magnetic phenomena in one neat set of formulae. 'Maxwell's equations' describe how a changing electric force 'field' creates a magnetic field, and how a changing magnetic field creates an electric field. The knitting together of electricity and magnetism into a seamless garment ranks as science's third great 'unification' after Newton's unification of Heaven and Earth, and Charles Darwin's unification of the human and animal worlds.[5]

Maxwell, on inspecting his elegant equations, noticed something very unexpected. They permitted a wave to ripple through the electric and magnetic fields that permeated empty space. And that was not all. The wave had a remarkable property: it propagated *at the speed of light in a vacuum*. The implication was as obvious to Maxwell as it was astonishing. Light must be an 'electromagnetic wave'. Not only had Maxwell found a link between electricity and magnetism, he had discovered a link between electricity, magnetism *and light*.[6]

Within two decades, Maxwell's theory had scored a remarkable technological success. The German physicist Heinrich Hertz, following the Scottish physicist's recipe, actually created artificial electromagnetic waves. In November 1886, using an electric spark as a 'transmitter', he broadcast invisible 'radio waves'.[7] They induced an electric current in a coil of wire, acting as a 'receiver', on the other side of his laboratory.

Our ultra-connected world, in which the invisible chatter of a billion voices courses through the air around us, was born on that day in 1886. 'From a long view of the history of mankind – seen from, say 10,000 years from now,' said the twentieth-century American physicist Richard Feynman, 'there can be little doubt that the most significant event of the nineteenth century will be judged as Maxwell's discovery of the laws of electrodynamics.'[8]

But, for all its triumphs, Maxwell's theory posed a very serious problem for physics. It was incompatible with the laws of motion of Galileo and Newton.

All waves ripple through something – water waves through water, sound waves through air. The hypothetical medium through which light rippled was christened the 'aether'.[9] An unavoidable consequence of the existence of this aether was that the speed anyone measured for a beam of light must depend on how fast they were travelling through the medium. Say you are standing on the deck of a sailing boat. The speed of the wind hitting your face depends on whether the boat is heading into the wind or whether it has the wind at its back. But the odd thing about Maxwell's equations was that they made no reference whatsoever to any light-carrying medium. Instead, they contained one, and only one, speed for a beam of light in a vacuum. It was immutable, constant, utterly impervious to the world in which it was embedded.

The obvious conclusion to draw from this was that Maxwell's equations were in error and required modification. They were, after all, the new kid on the scientific block. Newton's laws of motion, on the other hand, had been established almost two centuries earlier and, since that time, nobody had found a single instance in which reality departed from their predictions. Enter Einstein. He was mesmerised not only by Hertz's dramatic confirmation of Maxwell's equations but by their *beauty*, a quality he considered a strong indication of their *rightness*.

Newton had written in his notebook: 'Plato is my friend – Aristotle is my friend – but my greatest friend is truth.' Ironically, it was because Einstein concurred 100 per cent with this sentiment of his predecessor that he had the temerity to doubt him. And that was why, aged sixteen, he had asked himself the critical question: what would it be like to catch up a light beam?

Seeing the impossible

A light wave, according to Maxwell, is a complex beast consisting of an electric field and a magnetic field oscillating at right angles to each other and at right angles to the wave's direction of travel. The electric field grows as the magnetic field decays and vice versa. In fact, the decay of one field generates the other so that the two fields alternate to create a self-sustaining electromagnetic wave.

The details are not important here. It is sufficient to think of a light wave as like a water wave rippling across a lake. If you were to catch it up, it would appear stationary, a long train of undulations frozen as if in a photograph. But – and here is the problem that occurred to the teenage Einstein in Aarau, Switzerland – Maxwell's equations do not permit the existence of a stationary electromagnetic wave. Put simply, if you were to catch up a light beam, you would see something impossible – something that, according to the laws of physics, simply cannot exist.

How is it possible to resolve this paradox? If Maxwell's theory is correct, there is only one way, realised Einstein. Since travelling at the speed of light means seeing something impossible, travelling at the speed of light must itself be impossible. It is as simple as that. The trouble is that Newton's laws of motion permit a body to travel *at any speed whatsoever*. They say nothing about an ultimate cosmic speed limit.

Consequently, the price of saying that nothing material can travel at the speed of light is very high. It means overthrowing the worldview of Newton, the greatest scientist who ever lived. No one would do this lightly. Not without an awful lot of supporting evidence. And this is why Einstein spent nine years wrestling with the problem of squaring the theory of electromagnetism with the laws of motion. Everything finally came to a head in the spring of 1905.

Patent paradise

By now, Einstein, aged twenty-six, was a Technical Expert, Class III, at the Swiss Federal Patent Office in Bern, a post he had held since 1902. He was living in a two-room, third-floor apartment at Kramgasse 49 with his Serbian wife, Mileva Marić, and their one-year-old son, Hans Albert. Marić, four years his senior, had been the only woman in his class at the Swiss Federal Polytechnic in Zurich. Their romance had scandalised both their families, especially when a baby was born out of wedlock in 1902. Lieserl, the only reference to whom is in letters to and from Novi Sad, where Mileva returned to give birth, either died eighteen months later or was given up for adoption by Mileva's family. Einstein and Mileva, who hid Lieserl's existence from their friends in Switzerland, were the only ones who knew the truth of her fate.

The Patent Office saved Einstein's life, something he would

remain grateful for until the day he died. After failing to obtain either a teaching or a university post, and, by his own admission, half-starving, it also provided him with the income and respectability to marry Mileva in 1903. Although grief for the loss of Lieserl must have hung over their union, playing a part in dooming it from the start, Einstein's time at the Patent Office proved to be one of the happiest periods of his life.[10]

Not only did his job as a Technical Expert, Class III, pay the bills, it placed him right at the high-tech frontier of the new electrical age. His knowledge of electrical devices had been gained – sadly – at his father's failed electrical lighting company in Milan. But he put it to good use in his office on the top floor of the new Postal and Telegraph Administration building, near Bern's central train station, on Genfergass. Much to the approval of his boss, Friedrich Haller, Einstein was able to spot the subtlest flaws in the designs submitted each month to the Patent Office for dynamos, motors, transformers, and the like. But the best thing about his 48 hours a week as a Technical Expert, Class III, was that it did not overly tax his brain – as a teaching or university post might have done – and it left him time for creative thinking. And, boy, did he create.

The year 1905 is generally known in the annals of science as Einstein's 'miraculous year'. 'No one before or since has widened the horizons of physics in so short a time as Einstein did in 1905,' said physicist Abraham Pais.[11] No one, perhaps, except Isaac Newton. But, whereas Newton's 'miraculous year' lasted about eighteen months, Einstein's spanned barely more than three months. At least, that was the period – between 17 March and 30 June – when Einstein finished four scientific papers of such seismic importance that they would utterly remake the landscape of physics.

The first paper, which Einstein called 'very revolutionary' and which would earn him the 1921 Nobel Prize for Physics, questioned the very idea of light as a wave, and suggested that atoms instead spit out or gobble up light in tiny chunks, or 'quanta'.[12] The second paper, which would earn Einstein a doctorate at the University of Zurich, determined the true size of atoms – whose existence at the turn of the twentieth century was still far from universally accepted – from the way in which they diffused through a liquid.[13] The third paper pointed out that the curious dance of pollen grains suspended in water – so-called Brownian motion, first seen through a microscope by the botanist Robert Brown in 1827 – was the result of their jittery bombardment by water molecules.[14] Finally, the fourth in this remarkable series addressed the problem of the uncatchability of light.[15]

The catalyst was Michele Besso, whom Einstein visited in mid-May 1905. Besso, six years Einstein's senior, had been a close friend since 1896 when Einstein was studying for a teaching qualification at the Swiss Federal Polytechnic in Zurich and Besso was working as a mechanical engineer in nearby Winterthur. Both loved music – Einstein being a competent violin player – and they had met through a Zurich woman called Selina Caprotti who gave over her home on Saturday afternoons to people wanting to play music together.[16]

Besso not only recommended books for Einstein to read but entered into endless philosophical discussions with his friend about the foundations of physics. Most importantly, Besso was a critical sounding board for Einstein's ideas. Recalling his visit in mid-May to Besso to discuss the problem of the uncatchability of light, Einstein said: 'It was a beautiful day. We discussed every aspect of the problem . . .'[17] He did not say how long he and Besso talked, where they talked or whether the discussion was heated. But the outcome, according to Einstein, was like a light coming on in a dark room and in an instant revealing everything. 'Suddenly, I understood where the problem lay!'

Perhaps that evening Einstein discussed it with his wife, Mileva. Or perhaps he lay sleepless in bed, turning the problem over and over in his mind, examining it from every possible side in the manner of Newton. Or perhaps he worked at the kitchen table until the early hours, furiously filling page after page of his notebook with scribbled notes. There is no record. Mileva, preoccupied and worn down by her domestic chores, kept no diary, wrote nothing about these times, and was never at any later time interviewed by any journalist.

But, when Einstein saw Besso the next day, such was his state of excitement that he did not even say 'hello'. 'Thank you,' he said. 'I've completely solved the problem. An analysis of the concept of time was my solution. Time cannot be absolutely defined, and there is an inseparable relation between time and signal velocity.'[18]

Light plays the role of infinite speed

If a light beam is uncatchable, Einstein asked, what does that say about the speed of light? An analogy may help. Infinity is a number in mathematics that is bigger than any other number. If something were to travel at infinite speed, it would be impossible to catch. The fact that light is uncatchable must mean that, in our Universe, for some unknown reason, the speed of light plays the role of infinite speed. 'Nothing travels faster than the speed of light, with the possible exception of bad news, which obeys its own special laws,' said Douglas Adams.[19]

The analogy with infinite speed is useful. If something were moving at infinite speed, it would not matter what your own speed was or whether you were travelling towards it or away from it. Your speed would be so negligible by comparison that you would measure its speed to be infinite. Similarly, if the something moving at infinite speed were launched from a body travelling towards you or away from you, the body's speed would be so negligible by comparison, that once again the infinite-speed thing would always appear to be travelling at infinite speed. It follows that, if the speed of light plays the role of infinite speed, it always appears the same, irrespective of the speed of its source or of the speed of an observer. The speed of light is constant, doggedly unvarying for everyone no matter what their state of motion, exactly as Maxwell's theory suggests.

So much for generalities, what about the details? How, in practice, is it possible for everyone, no matter how fast they are moving, to measure exactly the same speed for a beam of light?

Well, speed is simply the distance a body travels in a given time – think of a car speeding along a motorway at 100 kilometres in an hour. If everyone is to agree on the same speed of light, something must therefore happen to each person's measurements of distance and time.

What actually happens, Einstein discovered, is that someone moving past you appears to shrink in the direction of their motion and, simultaneously, their time as shown by their watch appears to slow. Think of them flattening like a pancake and at the same time moving in slow motion.[20]

And all of this shrinking of space and slowing of time works in such a way that each person, no matter what their state of motion, estimates the distance that a light beam travels in a given time to be exactly the same. It is a huge cosmic conspiracy to ensure the constancy of the speed of light.

Of course, nobody ever sees space and time distort when someone walks by in a park or drives past in the street. This is because these strange effects would become apparent only if someone could fly past you at an appreciable fraction of the speed of light. But the speed of light is about a million times faster than a Boeing 747 and nothing in the everyday world even remotely approaches it.

Time dilation

Nevertheless, time dilation *is* detectable in the everyday world. Just. In 1971, super-accurate 'atomic clocks' were synchronised then separated, one being flown round the world on an airliner while the other stayed at home. When the clocks were reunited, the experimenters found that the round-the-world clock had registered the passage of marginally less time than its stay-at-home counterpart. The shorter time measured by the moving clock was precisely what is predicted by Einstein.

The slowing of time affects astronauts too. As the Russian physicist Igor Novikov points out: 'When the crew of the Soviet Salyut space station returned to Earth in 1988 after orbiting for a year at 8 kilometres a second, they stepped into the future by one hundredth of a second.'[21]

The time dilation effect is far greater for cosmic ray 'muons', subatomic particles created when cosmic rays – super-fast atomic nuclei from space – slam into air molecules at the top of the Earth's atmosphere. In fact, the evidence that time slows and space contracts at close to the speed of light is actually coursing through your body at this very instant.

Muons are created about 12.5 kilometres up in the atmosphere. They shower down through the air like subatomic rain. But here's the thing. A muon disintegrates after a characteristic interval of time. The interval is very short – a mere 1.5 millionths of a second. By rights, none should travel more than about 500 metres down through the atmosphere before disintegrating. Certainly, none should reach the ground, 12.5 kilometres below.

But they do.

The reason is that muons are travelling at 99.92 per cent of the speed of light. From your point of view, they live their lives in slow motion. In fact, time passes 25 times slower for them than for you, which means they take 25 times as long as expected to realise it is time to disintegrate. When they do, they have already reached the Earth's surface.

But, of course, there is another point of view – that of the muon. From its angle, time is passing at its normal rate – after all, a muon is stationary *with respect to itself*, as are you. Instead, it sees you shrink in the direction of its motion – or, rather, our motion, since, from the point of view of a muon, it is the ground that is approaching at 99.92 per cent of the speed of light. But not only do you shrink, so too does the atmosphere. It shrinks to a mere 1/25th of its normal thickness. Which means the muons have time to get to the surface before they disintegrate.

Whatever way you look at it – from your point of view, where the muon's time slows down; or from the muon's point of view, where the atmosphere shrinks – the muon gets to the ground. This is the magic of Einstein's theory.

'Aside from Velcro, time is the most mysterious substance in the Universe,' says American humorist Dave Berry. 'You can't see it or touch it, yet a plumber can charge you upwards of seventy-five dollars per hour for it, without necessarily fixing anything.'

No absolute space, no absolute time

The realisation that moving clocks slow – an effect known as 'time dilation' – and moving rulers shrink – known as 'Lorentz-FitzGerald contraction' – represents a seismic shift in our picture of reality.[22] It explains perhaps why all the great physicists of Einstein's time, despite being in possession of exactly the same facts, did not go where Einstein went. But then nobody except Einstein had the sheer bare-faced nerve to doubt Newton.

Newton, largely for pragmatic reasons, believed in 'absolute space', which existed as a backdrop to the Universe, like a great canvas on which the great cosmic drama was played out. Everyone would measure the same separation of two points in such space just as everyone would measure the same distance between two pins stuck in an artist's canvas.

But Einstein showed there is no such thing as absolute space.

Newton, in addition to absolute space, also believed in 'absolute time', which ticked away as if somewhere in the Universe there is a great master clock. Because of the existence of absolute time, everyone would agree on the same interval of time passing between any two events.

But Einstein showed that, just as there is no such thing as absolute space, neither is there is any such thing as absolute time. 'I can't talk to you in terms of time,' said the novelist Graham Greene, 'your time and my time are different.'

Exactly right. One person's interval of time is not the same as another person's interval of time, and one person's interval of space is not the same as another person's interval of space. Time and space are like shifting sand. The rock on which the Universe is built is the speed of light.

If all this seems a little imprecise, that is because it is. Einstein started his journey of discovery, aged sixteen, simply by imagining what it would be like to catch up with a light beam. This had revealed to him the shortcomings of the Newtonian view of motion and also hinted at what was needed to supplant it. But Einstein needed to construct a self-consistent theory, one founded on a minimum of assumptions, from which all the consequences for space and time would follow as inevitably as night follows day. How to do this was worked on by Einstein in the weeks following his pivotal meeting with Besso in May 1905.

The two foundation stones of relativity

Einstein built what would become known as the 'special theory of relativity' on two foundation stones.[23] The first is the assertion that the speed of light is independent of the speed of its source or the speed of an observer. And the second is the 'principle of relativity'.

Galileo, back in the seventeenth century, had realised that there was something odd about motion at constant speed in a straight line. *It does not change anything*. Say you throw a ball to a friend. It does not matter whether you are standing twenty paces away from them in a field or twenty paces away on the deck of a ship (providing the ship is moving smoothly and uniformly through the waves). In both cases, the ball loops through the air in precisely the same way.

From this common observation, Galileo concluded that the laws of motion are the same for all people who are moving at constant speed relative to each other. In other words, if you were to be beamed by a *Star Trek* matter transporter to a blacked-out cabin on-board a ship, you would not be able to tell from the flight of a ball through the air whether you were on-board a ship ploughing through the waves or stationary on dry land. In technical language, the laws of motion – distilled down to three basic edicts by Newton after Galileo's death – are 'invariant' with respect to motion at constant speed in a straight line. They will not reveal to you whether you are in such 'uniform motion'. And that is because the very idea of absolute motion – that is, motion with respect to absolute space *à la* Isaac Newton – is utterly meaningless.

Einstein extended 'Galilean relativity'. According to his principle of relativity, it is not just the laws of motion that are invariant with respect to uniform motion but *all the laws of physics*. In other words, there is no experiment you can do, including those involving the propagation of light, which can reveal whether you are moving or not.

As already mentioned, the aether was the hypothetical medium through which light rippled and against which it was possible to measure motion. Einstein's principle of relativity dispenses with the aether entirely.[24] The aether is revealed for what it is. A fantasy. An unfortunate cul-de-sac into which physicists had mistakenly wandered. No more than the nineteenth-century incarnation of Newton's 'absolute space'. Light needs no medium in which to propagate. It is a self-sustaining wave in the electromagnetic field.

With no fixed backdrop of absolute space with which to measure absolute velocity, the only meaningful concept is 'relative velocity'. When you see a person flying past you and their space shrinking and their time slowing compared to your own, you may wonder what you look like to them. The answer is *exactly the same as they look to you*. To them, you shrink in the direction of your motion and appear to be moving as if through treacle. There is complete symmetry because only relative motion matters. You are moving relative to them and they are also moving relative to you at the same speed (though in the opposite direction, of course). As the joke goes: 'When does Zurich stop at this train?' – Albert Einstein.

In summary, Einstein needed only two principles to build his revolutionary theory of space and time: the principle of relativity and the principle of the constancy of the speed of light.[25] And armed with these deceptively simple ideas, he was able to deduce absolutely everything else.

A fundamental basis for relativity

Einstein began by trying to define time pragmatically. 'Time', he said with child's simplicity and directness, 'is what a clock measures'.[26] The question then is: what is a clock?

Einstein imagined the simplest possible 'clock'. It consisted of a light source with a flat mirror some distance above it. The 'tick' of the clock was simply the time taken for the light to travel from the source up to the mirror, bounce off it and return to the source.

Think of such a clock in a train racing past you. In order to see the clock, of course, you would have to have X-ray eyes or the train would have to be transparent! But forget the details. This is just a 'thought experiment', designed to lay bare the fundamentals. The point is that, while the light is travelling up to the mirror, both it and the mirror move relative to you because the train is moving. So, instead of seeing the light travel vertically upwards to the mirror, you see it travel at a slant towards the mirror. And, similarly, when the light bounces off the mirror, you see it travel back down to its source at a slant. From your point of view beside the train track, the light does not simply travel up and down, it travels along two sides of an isosceles triangle. Since it has to cover a larger distance, the tick of the moving clock takes longer, demonstrating in a nuts-and-bolts way that moving clocks do indeed run slow.

A similar geometrical argument can be used to show that, from your point of view standing beside the train track, a ruler on the moving train shrinks in the direction of motion.

If you think all this reasoning concerns rather artificial abstract clocks and rulers, it turns out that the very atoms you are made of function as tiny clocks and rulers. Einstein's logic is inescapable. There is absolutely no way of sidestepping it. All clocks – which in the final analysis have to be read by means of light bouncing off them – ultimately boil down to the simple clock just described.[27]

Time slows and space shrinks everywhere, depending on your state of relative motion. One person's time is not another person's time.[28] One person's space is not another person's space. The measurement of time and space are inextricably bound up with signal velocity – the speed of light. And, because they are, reality is profoundly affected by its dogged constancy.

It took Einstein five weeks to write his paper. In the process he overthrew the Newtonian world view and replaced it with his own. To his colleague Josef Sauter at the Patent Office, he said: 'My joy is indescribable.'[29]

'On the electrodynamics of moving bodies' was published on 28 September 1905. Usually, at the end a scientific paper, the author lists the papers of other scientists which have influenced the work. Einstein acknowledged no other papers. In fact, the only other scientists he mentioned were the greats such as Newton and Galileo, Clerk-Maxwell and Hertz, and he used their names merely as labels for their theories. But then no one had influenced Einstein's thinking. Not fundamentally. Many had seen fragments of the new picture of reality. But no one else had seen it in its entirety – the fundamental unifying principles that tied everything together.

It was rather like Halley, Wren and Hooke all guessing the inverse-square law of gravity. But, without the vision – the bird's-eye view afforded Newton by his precise definitions of mass and force, and by his fundamental 'laws of motion' – the insight amounted to nothing. Newton alone had the vision. And that was why he, like Einstein, was the one to change our fundamental worldview.

But Einstein's paper did not simply lack a reference list of other scientific work. Usually, an author thanks all the people who have helped with advice or in discussing the author's work. But Einstein was the ultimate outsider, working in splendid isolation at the Swiss Patent Office in Bern, unknown in scientific circles. At the end of his paper, he thanked just one person: 'My friend and colleague Michele Besso steadfastly stood by me in my work on the problem here discussed,' he wrote. 'I am indebted to him for many a valuable suggestion.'

Space-time

The consequences of the speed of light being the rock on which the Universe is founded are more than simply that one person's time is not the same as another's time, and one person's space is not the same as another's space. It is worse than this. It turns out that one person's space is another person's space *and time*, and one person's time is another person's time *and space*.

None of this is apparent in the cosmic slow lane of the every-day world. But it would be glaringly obvious if you could travel at close to the speed of light. Space and time are not only like elastic, able to stretch without limit, but they can morph one into another. Ultimately, the reason for this is that they are aspects of the same thing: space-time.

We are used to thinking of there being three dimensions of space – east–west, north–south, up–down – and one dimension of time – past–future. But, actually, the dimensions of space and time intermingle to create four dimensions of space-time. Being denizens of a 3D world, we cannot perceive 4D space-time in its entirety. Instead, living in nature's slow lane, we perceive only 'shadows' of the 4D reality cast on our 3D world: one shadow being time, the other three being space.

Einstein, while a student at the Swiss Federal Polytechnic in Zurich, was taught by a mathematics professor called Hermann Minkowski. He famously referred to his student as a 'lazy dog'. Later, much to his credit, Minkowski recognised the genius of Einstein, and in fact recognised something that his student had not spotted in his own theory: that it unifies space and time. 'From now on, space of itself and time of itself will sink into mere shadows and only a kind of union between them will survive,' said Minkowski.

'The most important single lesson of relativity theory,' said Stephen Hawking's collaborator, the British mathematician Roger Penrose, 'is, perhaps, that space and time are not concepts that can be considered independently of one another but must be combined together to give a 4-dimensional picture of phenomena: the description in terms of space-time.'[30]

That there is such a thing as space-time, so that time shares some of the properties of space, means that the events of the Universe can be imagined spread out across a 4D map exactly like geographical features on a standard 2D map. From our perspective, within the map, time appears to flow. But, from an Einsteinian, 'bird's-eye' perspective, time does not flow. All events – from the big bang to the end of the Universe – exist simultaneously, laid out on the 4D map of space-time. Each person's life is a chain of events, referred to by physicists as a 'world line', which stretches like a snake across the map.

'The objective world does not happen, it simply is,' wrote the German physicist Hermann Weyl in 1949. 'Only to the gaze of my consciousness, crawling upward along the world line of my body, does a section of this world come to life as a fleeting image in space which continuously changes in time.' Weyl implicitly recognised that our experience of time flowing has no explanation in physics but only in biology, and in the way the human brain processes reality.[31] 'Reality is merely an illusion, albeit a very persistent one,' said Einstein.

The idea of all events existing simultaneously on the 4D map of space-time proved a source of comfort to Einstein when his good friend Besso died in 1955. 'Now he has departed from this strange world a little ahead of me,' he wrote to his bereaved family. 'That means nothing. People like us, who believe in physics, know that the distinction between past, present and future is only a stubbornly persistent illusion.'

Mass and energy

Space and time are the foundation stones of pretty much all other physical concepts. So, when they are shown to be nothing but shifting sand, so too are lots of other things in physics. Take electric and magnetic fields. Just as space and time are aspects of the same thing – space-time – electric and magnetic fields turn out to be aspects of the same entity – the electromagnetic field. In fact, this insight of Einstein's resolves a paradox in Maxwell's theory.

According to Maxwell, if you travel alongside an electric charge such as an electron, so that it is not moving relative to you, you feel an electric force field. If the electric charge is moving relative to you, however, you feel an electric field *and a magnetic field*. Similarly, if you travel alongside a magnet, you feel a magnetic field. But, if the magnet is moving relative to you, you feel a magnetic field *and an electric field*.

How is it possible that, from one perspective, there is an

electric field and, from another, no electric field? How is it possible that, from one perspective, there is a magnetic field and, from another, no magnetic field? The answer, Einstein realised, is that an electric field and a magnetic field are simply different facets of the same thing – an electromagnetic field – and how much of each facet you see depends on your speed relative to the source of the electromagnetic field.

But Einstein showed not only that electric and magnetic fields are two sides of the same coin, and that space and time are facets of the same basic entity, he also showed that mass and energy are aspects of the same thing.[32] And this last unification was maybe the greatest of all the consequences of the special theory of relativity.

$$E = mc^2$$

By the time Einstein's paper on the foundation of relativity was announced to the world in the 28 September 1905 issue of *Annalen der Physik*, its editor had received a three-page supplement from Einstein. It contained perhaps the most famous equation in all of physics: $E = mc^2$.[33]

It was an extraordinary and unexpected result. Mass, it turns out, is merely another form of energy, like sound energy or heat energy or electrical energy. Its distinguishing feature is merely that it is the most compact form of energy. In fact Einstein's formula, which multiplies the mass, m, of a body by the square of a very big number – the speed of light, universally referred to by physicists by the letter c – reveals that even the tiniest of masses contains a mind-bogglingly huge amount of energy, E.

It is a fundamental feature of the world that one form of energy can be converted into another form – for instance, electrical energy can be transformed into light energy in a light bulb and the chemical energy of food can be converted into the energy of motion of your muscles. Mass-energy is no exception. It too can be converted into other forms of energy such as heat and light. The appalling reality of that would be demonstrated to the world in August 1945 over the Japanese cities of Hiroshima and Nagasaki.

But Einstein's $E = mc^2$ formula can be read both ways. Not only is mass a form of energy but energy has an effective mass. Any type of energy. So sound energy has a mass, heat energy has a mass, chemical energy has a mass, and, crucially, so does energy of motion.

So a body has an intrinsic mass – universally known as its 'rest mass' – but it also possesses mass due to its motion. In other words, as a body is boosted in speed, it becomes more massive. You weigh more if you are running for a bus than if you are standing still at a bus stop. A mug of coffee weighs more when it is hot than when it is cold because 'temperature' is a measure of microscopic motion, and the molecules in the coffee jiggle about more rapidly when it is hot than when it is cold. Of course, such increases in mass become appreciable only when a body is close to the speed of light, making them too small to notice in everyday circumstances.

But, as a body is boosted in speed and becomes more massive, it becomes harder to push. In fact, if a material body were ever to attain the speed of light it would become infinitely massive, which is impossible. There is simply not enough energy in the Universe. This then is an explanation for why a light beam is uncatchable.[34] Everything hangs together. Einstein's special theory of relativity is a beautiful, seamless whole.

For light, which has no rest mass and which can travel at the cosmic speed limit, time slows to a standstill, and its birth at the beginning of the Universe and its death at the end of the Universe are simultaneous events. 'What binds us to space-time is our rest mass, which prevents us from flying at the speed of light, when time stops and space loses meaning,' says the Ukrainian mathematician Yuri Ivanovitch Manin. 'In a world of light there are neither points nor moments of time; beings woven from light would live "nowhere" and "nowhen"; only poetry and mathematics are capable of speaking meaningfully about such things.'

Generalising relativity

The special theory of relativity swept aside Newton's concepts of absolute space and absolute time, revealing Newtonian physics to be wrong, although still a fantastically good description of the everyday world. But, despite the theory's tremendous success in radically transforming our picture of reality, it had several problems.

The first was that it spelled out what must be done to the measurements of space and time of people moving at constant speed relative to each other so that they agree on the same laws of physics – that is, the same laws of motion and the same laws of optics, principally the law of the constancy of the speed of light. But people moving at constant speed relative to each other are not typical. In the real world, observers change their velocity. A car slows to a halt on a red traffic light, then speeds up again on green. A rocket gets ever faster until it attains the speed required to orbit the Earth.

The task facing Einstein was clear. He needed to find what must be done to the measurements of space and time of people varying their speed, or 'accelerating', relative to each other so that they would agree on the same laws of physics. Those laws should look the same no matter how people are moving – falling, spinning or being pressed into the seat of an accelerating car. He needed to turn his 'special' theory into a 'general' theory of relativity.[35]

There was nothing mysterious about Einstein's desire. If the laws of physics are to have the status of universal laws they should be independent of our point of view. It should not matter, for instance, whether we are sitting next to a bar magnet, moving past it at constant speed or accelerating past it. We should see the same fundamental law of magnetism.

But special relativity had other problems in addition to not dealing with accelerated motion. Most seriously, it was in fundamental conflict with Newton's theory of gravity.

All Newton's law of gravity did was specify the strength of the gravitational force at every distance from a massive body like the Sun. This is tantamount to saying that the gravity of a massive body is felt everywhere *instantaneously*, and this in turn is equivalent to saying that the effect of gravity propagates at infinite speed. But, according to special relativity, nothing, not even gravity, can surpass the cosmic speed limit set by the speed of light.

The predictions of Newton's theory of gravity and Einstein's special theory of relativity would be most at odds if the Sun were to vanish. Obviously, this is an unlikely event! If it did happen, the Earth would notice immediately according to Newton, and fly off on a tangent towards the stars. But, according to Einstein, the planet would continue merrily in its orbit for the time it takes light to travel between the Sun and Earth. Only after an interval of 8½ minutes would it notice that the Sun had gone and head for the stars.

The way to incorporate the cosmic speed limit set by light into a theory of gravity, Einstein realised, was to use the concept of a 'field'. This had been invented by the English scientist and electrical pioneer Michael Faraday in the early nineteenth century.[36] Faraday had a strong sense, when he held a piece of iron in the vicinity of a magnet and felt it gripped by a powerful force, that there was an invisible force-field extending outwards from the magnet. In fact, when he sprinkled iron filings around the magnet he was able to make visible the 'lines of force'.

In Faraday's view, a magnet does not exert a force directly on a piece of iron. Instead, a magnet sets up a magnetic field of force around it, like a *Star Trek* tractor beam, and it is the field that exerts a force on the iron. It might seem like a subtle difference. But not only does this picture give the field a physical existence – in the case of the electromagnetic field, a vibration passing through it is a physical electromagnetic wave (light) – but it admits the possibility of the field propagating outwards at some speed.[37]

By analogy with electromagnetism, Einstein needed to create a theory in which a mass produces a gravitational field and the gravitational field in turn exerts a force on other masses.

Crucially, it is possible for such a field to propagate at a particular speed and so to incorporate the cosmic speed limit of the speed of light.

But creating a field theory of gravity that is compatible with special relativity was only the second of Einstein's problems. A third problem arose because the 'source' of gravity in Newton's theory is mass. But Einstein had discovered that all forms of energy possess an effective mass, and so exert gravity. Consequently, the ultimate source of gravity cannot be mass. It must be energy.

Einstein had almost certainly been aware of all of these problems with special relativity since its completion in 1905. But things came to a head in October 1907. It was then that he was invited by the German physicist Johannes Stark to write a comprehensive summary of special relativity in *The Yearbook of Radioactivity and Electronics*.

Einstein was still working at the Swiss Federal Patent Office in Bern, though since 1 April 1906 he had the elevated rank of Technical Expert, Class II. Working after office hours, he completed the review article in two months, delivering it to Stark on 1 December 1907. The first four of its five sections set out the basic ideas of special relativity and worked through their consequences for space, time, matter and energy. The fifth section was entitled 'The relativity principle and gravitation'.

While other physicists were still struggling with the counter-intuitive ideas of special relativity, Einstein had already realised that the theory was nothing more than a beginning. In a letter to his friend Conrad Habicht at the end of December, he said he was pursuing another relativity theory. 'But so far it does not seem to work out,' he admitted in a postscript.[38]

Those were prescient words. It would take him another eight years to achieve his goal of extending the relativity principle to gravity and obtaining a 'general' theory of relativity. And it might have taken him even longer had it not been for a crucial insight he had while staring out of a window at the Patent Office.

Further reading

Bais, Sander, *Very Special Relativity*, Harvard University Press, Cambridge, MA, 2007.

Einstein, Albert, *Relativity: The Special and General Theory*, Folio Society, London, 2004.

Fölsing, Albrecht, *Albert Einstein*, Penguin, London, 1998.

Jaffe, Bernard, *Michelson and the Speed of Light*, Anchor Books, Garden City, NY, 1960.

Overbye, Dennis, *Einstein in Love: A Scientific Romance*, Viking, London, 2000.

Pais, Abraham, *'Subtle is the Lord . . .': The Science and the Life of Albert Einstein*, Oxford University Press, Oxford, 1983.

Ode to a falling man

How Einstein realised that the 'force' of gravity is an illusion and all there really is is warped space-time

If a bird-watching physicist falls off a cliff, he doesn't worry about his binoculars; they fall with him.

Sir Hermann Bondi[1]

In some sense, gravity does not exist; what moves the planets and the stars is the distortion of space and time.

Michio Kaku[2]

A falling person does not feel their weight. This insight, which occurred to Einstein in 1907, would become the foundation stone on which he would build the edifice of a new and revolutionary theory of gravity. But frustratingly, just as with his insight that catching up a light beam is impossible, Einstein did not reveal the precise circumstances that triggered the revelation. Instead we can only speculate. We know that, at the time, Einstein was living and working in the Swiss capital. 'The breakthrough came suddenly one day,' he wrote. 'I was sitting on a chair in my patent office in Bern.'

I imagine Einstein at his desk, considering the last patent application of the day, which he has read all the way to the bitter end:

47242
Allgemeine Elektricitätgesellschaft, Berlin
Nägeli & Co., Bern
Alternating current machine

He dabs the nib of his fountain pen on the blotter before taking a fresh piece of Swiss Federal Patent Office paper from the stationery tray. He hesitates no more than a second or two to compose his thoughts. Then, swiftly (and somewhat devastatingly), he writes: 'Point 1: The patent claim is incorrectly, imprecisely, and unclearly prepared.'[3]

He does not get to 'Point 2'.

The scream jolts him like an electric shock. Jumping to his feet, he sees the roofer skittering down the tiled roof of the building across the street. The man is flailing his arms desperately and picking up speed inexorably. But, before he flies off the edge of the roof and plummets five storeys to the busy Genfergass – at the very last instant possible – he lunges for a flagpole. It appears too flimsy to hold him. But – miracle of miracles – it bends but does not snap.

I picture Einstein watching the whole drama unfold from the Patent Office on the top floor of Bern's new Postal and Telegraph Administration. Only when he sees the roofer being hauled back to safety by his workmates, does he sit back down, relieved, at his desk. With his heart still hammering, it is a while before he can focus again on Patent Application 47242.

Is his comment too harsh? Has he allowed himself to be influenced by his father's bitter experience in Munich, where Elektrotechnische Fabrik J. Einstein & Cie had competed, unsuccessfully, with a number of highly aggressive companies – among them AEG – for the contract to light the city centre? No, he is confident he is simply being honest not vindictive. But in 'Point 2' he is careful to spell out in more conciliatory language his specific objections to Patent Application 47242. He then blots his page, leans back in his chair and looks at his now-empty in-tray with satisfaction.

His boss and saviour, Friedrich Haller, is off in Zurich on business, and his office-mate and friend, Josef Sauter, has taken advantage of Haller's absence to go to the Bear Pits, where he thinks he left his favourite umbrella at the weekend, and to buy his wife an anniversary present. (With a pang of guilt, Einstein realises he has never bought Mileva an anniversary present.)

The office is empty and quiet. I see Einstein leaning back in his chair to think. He recalls the dramatic events across the street and replays the alternatives in his head. The roofer skitters down the roof and grabs the flagpole which bends but holds his weight. The roofer skitters down the roof, grabs the flagpole but it bends and snaps and he sails out into space.

He imagines what it would be like to fall down to the street and his stomach lurches. He grips the desk. He catches his breath. In such circumstances, he has heard it said, subjective time slows almost to a standstill and a whole lifetime of events parades past one's eyes. But what if it were possible to fall for ever?

He imagines falling in a place where there is no air or wind to slow him down. He is falling through time and space and stars and sky and everything in between. He is falling until he forgets he is even falling.[4]

And then it hits him like a bolt of lightning!

He leaps up abruptly, knocking his chair backwards. Instantly, he knows he has stumbled on the foundation stone on which he can build a new reality. In later years, he will call it the happiest thought of his life. It is so obvious that in the empty office he laughs out loud. Of course!

A falling person does not feel their weight!

Did a roofer fall down a roof and furnish Einstein with his moment of inspiration? Or did some other event, something rather less dramatic, trigger his light-bulb moment? Though it is fun to imagine, we will of course never know. All that Einstein told us is that some time in 1907, he had the seemingly innocuous thought that set him on the road to overthrowing the worldview of Newton.

But why is the realisation that a falling person does not feel their weight such a key insight? Picture the following situation.

A man is travelling in a lift when – disaster – the cable snaps.[5] Instantly, he finds himself in free-fall. Say, he is standing on a set of scales on the floor of the lift. (This is not a very realistic scenario!) One moment, the scales read 70 kilograms, the next exactly zero. This, in concrete terms, is what it means to not feel your weight when you are falling.

According to Newton, it is impossible to get beyond the pull of gravity since it merely weakens with distance but never completely disappears. According to Einstein, however, it is easy to get beyond gravity. All you have to do is free-fall. Gravity vanishes and you become weightless.

The situation of a falling man who feels no gravity is indistinguishable from that of a man floating in empty space far from the gravity of any planet. This provides a bridge between a theory of gravity and special relativity since, in both situations, the laws of special relativity apply.

The weighing scales read zero when the man is in free-fall because, as fast as he plunges down towards the scales, the scales plunge downwards away from him. In other words, the man falls at exactly the same rate as the scales, despite the man having a mass of 70 kilograms and the scales considerably less.

That all things – not just a 70-kilogram man and a set of scales – fall at the same rate under gravity was first noted by Galileo in the seventeenth century. According to legend, he dropped a heavy mass and a light mass from the Leaning Tower of Pisa and saw that they hit the ground at the same instant.

On Earth, such an experiment is inevitably complicated by the effect of air resistance, which creates a greater drag on a body with a larger surface area. But, in 1972, Apollo 15 commander Dave Scott repeated Galileo's experiment on the Moon where, of course, there is no air. He dropped a hammer and a feather from the same height. And, sure enough, two simultaneous puffs of lunar dust confirmed that the two objects struck the ground at the same time.

That all bodies fall at the same rate under gravity irrespective of their mass is actually extremely odd. Think of applying the same force to a big mass and a small mass – say, a loaded fridge and a wooden stool. Everyday experience tells us that the fridge will speed up, or 'accelerate', the least because bigger masses are harder to budge than smaller masses.[6] They have greater reluctance to move, or 'inertia'. In fact, this reluctance is the very basis of our concept of 'mass'.

But the weird thing in the case of masses experiencing the force of gravity is that, even though a bigger mass is harder to budge than a smaller mass, the force of gravity appears to adjust itself so it is greater on the bigger mass – and by exactly the amount necessary for it to fall at the same rate as the smaller mass. So a body that has twice the mass of another experiences twice the gravitational force, a mass that is three times as massive, three times the force, and so on. Drop a loaded fridge and a stool from the Leaning Tower of Pisa, or, better still – not only for safety reasons (!) but to avoid air resistance – drop them on the Moon, and they will hit the moon dust just as simultaneously as Dave Scott's hammer and feather.

Technically, the resistance of a body to being budged depends on its 'inertial mass', m_i (as embodied in Newton's second law, which says that, when a body is subjected to a force, F, it accelerates by an amount F/m_i). And, technically, the force applied by gravity to a body depends on its 'gravitational mass', m_g.

A body with twice the inertial mass of another has twice the resistance to being budged. It falls at exactly the same rate as a smaller mass only because it also experiences twice the force of gravity. In other words, a body's resistance to motion, which depends on its inertial mass, goes up exactly in step with the force of gravity, which depends on its gravitational mass. This is tantamount to saying that gravitational mass, m_g, and inertial mass, m_i, are identical.

Everyone since Galileo believed that a body's resistance to being budged and the force it experienced from gravity are two entirely different things. They certainly appear unconnected. It was Einstein's genius to realise that everyone since Galileo was wrong and that they had entirely missed what was staring them in the face. That a falling person does not feel their weight – or, equivalently, that all bodies accelerate at the same rate under gravity – can mean only one thing. Gravitational mass and inertial mass are the same. In other words, gravity *is* acceleration.

As already pointed out, in 1907 Einstein knew that he needed to generalise his theory of relativity so that it described the world from the point of view of not only people moving at constant speed relative to another but also accelerating with respect to each other. He also knew he needed a new theory of gravity since Newton's theory of gravity is incompatible with special relativity. How remarkable, then, to discover that a generalised theory of relativity is automatically a theory of gravity. It is the ultimate 'buy one, get one free'.

The power and simplicity of Einstein's key insight takes a little thought to appreciate. If gravity and acceleration are identical, then there is no need for gravity to adjust itself so that all bodies, no matter what their mass, fall at the same rate. It happens entirely naturally and automatically. This is how . . .

Rocket man

Say an astronaut wakes up on a rocket far from the gravity of the Earth or of any planet. The rocket is accelerating at 1g so his feet are glued to the floor of his cabin and he can walk about inside just as if he is on the Earth's surface.[7] In fact, if the windows of the rocket are all blacked out, he may think he is in a room on the surface of the Earth. Einstein went further than this. His contention was that there is no way the astronaut can prove that he is not on the surface of the Earth. In practice, gravity is indistinguishable from acceleration.

Now say the astronaut – out of boredom or curiosity – tries to recreate the experiment of Galileo and Dave Scott. At shoulder height he holds out a hammer and a feather, before releasing them together. They appear to fall at the same rate and hit the floor of the cabin simultaneously. Of course, the astronaut, who has no idea he is in a rocket and thinks he is on the surface of the Earth, attributes this to gravity, which causes all bodies to fall at the same rate.

But we know better. We are certain he is not on the surface of the Earth but far away from the gravity of any planet. When he let go of the hammer and the feather what really happened is they hung motionless in space, and the floor of the cabin *accelerated up towards them at 1g*. It hit the hammer and the feather simultaneously. After all, how could it not?

What this shows is that, if gravity is acceleration, the explanation of why all masses fall at the same rate is utterly trivial. There is no need for gravity to adjust itself to each mass to make this happen. No wonder Einstein called it the happiest thought of his life.

Gravity, Einstein realised, is not like other forces. It is an illusion. It is caused by us accelerating and not realising it. Einstein codified the idea that 'gravity is indistinguishable from acceleration' in his Principle of Equivalence. It became the foundation stone of his theory of gravity.

But why do we mistake acceleration for gravity? Because, as Einstein perceived, we have a limited perspective. And our limited perspective, like that of the astronaut in the blacked-out rocket, does not permit us to see the reality of our situation. That reality is that we are living in warped space-time. This takes a little explanation.

Linear acceleration implies warped space

The astronaut in the blacked-out rocket – out of boredom or curiosity again – carries out another experiment. This time it involves a laser. He takes the laser and puts it on a shelf 1 metre above the floor of his cabin. He switches it on so that its beam stabs horizontally across the cabin, creating a bright-blue spot on the far wall. He walks across and is puzzled to see that the spot is less than 1 metre above the floor. As it crossed the cabin, the light beam appears to have curved downwards.[8]

We of course know that the rocket is accelerating at 1g. So, while the light beam crosses the cabin, the floor accelerates up towards it. It is no surprise to us that the beam strikes the far wall of the cabin at a point less than 1 metre above the floor. The puzzled astronaut, however, believing he is experiencing gravity on the surface of the Earth, concludes that the path of light is bent in the presence of gravity, or, equivalently, *gravity bends light*.

But why does gravity bend light? One of the defining features of light is that it always takes the shortest path between two points. Usually, the shortest path is a straight line. But not always, realised Einstein.

Consider a hiker negotiating a wild and rugged landscape between two hilltops. Being an experienced hill walker, he finds the shortest path. Now say a woman in a microlight is flying high above the landscape. She is able to follow the hiker's path because of his hi-vis outfit. And the path she sees from her eagle's vantage point is a winding and tortuous one.

What this illustrates is that, in a hilly landscape, the shortest path between two points is not a straight line. It is a winding and tortuous path. A curve.

This has implications for the astronaut who sees his laser beam curve downwards as it crosses his cabin. The only way the shortest path can be a curve is if the space in the cabin is warped, like the hiker's hilly landscape.

Gravity therefore bends light because gravity is synonymous with warped space. *It is warped space.* It is hard to imagine a more profound shift in our picture of gravity from Newton's view.

Rotational acceleration implies warped space

The example of the rocket illustrates acceleration in a straight line. But it turns out that any acceleration is associated with warped space. Imagine, for instance, a spinning roundabout.

Any body that changes its velocity – that is, its speed or direction or both – is said to be 'accelerating'. The roundabout is doing just this. Although the natural motion of every element of the roundabout is to travel at constant speed in a straight line, every piece is constantly being dragged away from its desired straight-line path and made to move in a circle.

Think of laying 1-metre rulers, end to end, around the periphery of the roundabout and across the diameter of the roundabout. If the roundabout is 5 metres in diameter, five metre rulers will be needed to span it and about sixteen metre rulers to stretch around the circumference. This is because, as every schoolchild knows, the circumference of a circle of diameter, d, is given by $\pi \times d$.

Now imagine that the roundabout is spinning not just fast but super-fast – so fast that all points along the periphery are travelling at an appreciable fraction of the speed of light. According to Einstein's special theory of relativity, the rulers shrink in their direction of motion. It may now take twenty rulers or fifty or a hundred to stretch around the circumference, depending on how fast the roundabout is spinning. Contrast this with the rulers spanning the diameter of the roundabout. They are moving perpendicular to their length, not in the direction of their length. So they suffer no relativistic shrinkage, and five 1-metre rulers are still sufficient to span the diameter of the roundabout.

So how do we explain that the circumference of the roundabout is now much greater than $\pi \times d$? A hint comes from the circumference of a circle being $\pi \times d$ only on a flat surface like a piece of paper.

Think instead of a circle drawn on a sphere. Its circumference is less than $\pi \times d$. By contrast, a circle drawn on something that curves the opposite way – say, a deep valley in a trampoline – has a circumference greater than $\pi \times d$. This suggests an explanation of why the circumference of the roundabout is greater than $\pi \times d$: the space occupied by the roundabout is curved, or warped.

So, whatever type of acceleration is considered – whether in a straight line or in a circle – the result is the same. Acceleration is associated with curved space. And since gravity *is* curved space, the acceleration associated with rotation can simulate gravity. This was famously depicted in the movie *2001: A Space Odyssey*, where a space station in Earth orbit spins like a giant wheel and those inside can walk around the periphery of the wheel with their feet pinned to the floor by its artificial gravity.

Actually, gravity is more than simply warped space.

In the case of special relativity, one person's space turned out to be another person's space and time; one person's time another person's time and space. It is this realisation that led Herman Minkowski to his key insight that space and time are aspects of a seamless entity: space-time. Gravity, then, is not merely warped space, it is warped *space-time*.

Minkowski's concept of space-time, which Einstein, for all his genius, had not anticipated, proves absolutely crucial to understanding gravity.

Warped time

Since gravity is warped space-time, it follows that it not only plays games with space – bending the paths of light beams – but it also plays havoc with time.

Picture an idealised 'clock' that consists of a horizontal laser beam bouncing back and forth between mirrors. Each time the light strikes a mirror, it is detected, creating a 'tick'. If the clock is on the Earth, then the light does not travel between the mirrors in a perfectly straight line but follows a curved path – because, of course, gravity bends light.

Now think of two such clocks – one higher above the ground than the other. The lower clock is in slightly stronger gravity than the higher clock since it is closer to the bulk of the Earth. This means that the light travelling between the mirrors in the lower clock follows a more curved path than the light in the higher clock. The more curved the path, the further the light has to travel between the mirrors, and the greater the time between ticks. It follows that the lower clock ticks more slowly than the upper clock. In other words, *time flows more slowly in strong gravity*.[9]

Incredibly, this means that you age more slowly on the ground floor of a building than on the top floor. The reason is that, on the ground floor, you are closer to the mass of the Earth so gravity is marginally stronger. In fact, in 2010, physicists at the National Institute of Standards and Technology in the US were able to show that, if you stand on one step of a staircase, you age more slowly than someone standing on the step above you.[10] It is an extremely tiny effect – because the gravity of the Earth is relatively weak – but it is measurable with two super-sensitive atomic clocks.

If you think this is an esoteric effect with no relevance to everyday life, think again. SatNavs and Smartphones use data from a constellation of Global Positioning Satellites, which swing round the Earth in highly elongated orbits. GPS satellites carry on-board clocks, and, when they swoop in close to the planet, they experience stronger gravity and their clocks slow down. If your electronic devices did not compensate for this effect of general relativity they would be unable to pinpoint your location relative to the GPS satellites.

In other words, many of us on a daily basis are inadvertently taking part in an experiment that tests the general theory of relativity. If the theory were false, then the GPS system would get your location wrong by about an extra 50 metres a day. In fact, after ten years, it will still be correct within 5 metres, showing just how accurate is general relativity.[11]

There is another way in which the slowing of time in gravity manifests itself. Say a person is in a room on Earth rather than on a spaceship. They shine a blue laser beam up at the ceiling. They discover something odd. The spot on the ceiling is not blue. It is red. The reason is that the light originated closer to the mass of the Earth, where gravity is stronger and clocks tick more sluggishly. The up-and-down oscillation of a light wave is just like the tick of a clock, which means that the light also oscillates more sluggishly. Since 'colour' is just a measure of how fast light is oscillating, with red light vibrating less than blue light, the more-slowly-ticking light is red.

On the Earth, this 'gravitational red shift' of light climbing upwards is extremely small. It is certainly not enough to change the colour of light from blue to red – I was exaggerating. The colour shift is nevertheless measureable in super-accurate experiments. In one such experiment, in 1959, the American scientists Robert Pound and Glen Rebka observed the gravitational red shift of light climbing up a 22.6-metre tower. It was a tour de force because the shift is so tiny over such a short distance. But the effect is relatively easy to see in the light of 'white dwarfs', highly compacted stars which have very strong surface gravity.

Gravity affects time because gravity is not simply warped space but warped space-time. The warped-space part bends the path of light. And the warped-time part slows clocks.

Warped space-time

It took Einstein to realise that we are living in warped space-time and that warped space-time is, in fact, gravity. Nobody else noticed because it is far from obvious.

Imagine a race of intelligent ants that live on the surface of a trampoline and are confined to its two-dimensional surface. They can wander north and south, and east and west, but they have no perception of a third dimension – that is, above and below the trampoline. Now imagine someone puts a bowling ball on their trampoline. The ants discover that, as they cross from one side to the other, their paths are deflected toward the bowling ball. This peculiar deflection cries out for an explanation. And the ants find one. They reason that the bowling ball exerts a force of attraction on them. Perhaps they even christen the force 'gravity'.

But, looking down on the trampoline from the God-like perspective of the third dimension, things look different. It is abundantly clear that the bowling ball has created a depression, or valley, around it. And in taking the shortest path across the trampoline, the ants naturally follow a path that curves around the bowling ball, just as the hiker followed a curved path when negotiating the hilly landscape.[12]

We are in much the same situation as the ants on the trampoline. We live in a 3D world and are unable to perceive the full 4D reality in which it is embedded. The Sun creates a valley in the 4D space-time around it just as surely as the bowling ball created a valley in the 2D trampoline. Because we do not see it, we attribute the fact that the Earth's path through space is deflected in a circle around the Sun – or, strictly speaking, an ellipse – to a 'force' which reaches out from the Sun and grabs hold of the Earth. But, in reality, there is no such force – no invisible elastic binding the Earth and Sun – just as there is no force reaching out from the bowling ball.

The natural motion of a body subject to no other force is to follow the straightest possible trajectory through curved space-time. Consequently, the Earth circles the Sun like a roulette ball in a roulette wheel. 'In some sense, gravity does not exist,' says American physicist Michio Kaku. 'What moves the planets and the stars is the distortion of space and time.'[13]

This reveals the very essence of Einstein's theory of gravity. The American physicist John Wheeler put it this way: 'Matter tells space-time how to curve. And curved space-time tells matter how to move.' It is as simple as that. Actually, it is energy that warps space-time – mass-energy being just one form of energy. But this is just nit-picking. Wheeler's statement is a masterful distillation of the essence of the general theory of relativity.

To bring the idea down to Earth – literally – there is a valley in the space-time around the Earth. Our natural motion is to fall to the bottom of the valley – that is, to the centre of the Earth.[14] But the surface of the Earth gets in the way. It obstructs our natural motion. The upward force from the ground is how we experience gravity.

The contrast between Newton's and Einstein's theories of gravity is striking. In Newton's theory, the Earth wants to move uniformly in a straight line because this is what massive bodies naturally do. But the gravitational force from the Sun deflects the Earth from its desired 'inertial' motion and causes it to travel in an elliptical orbit around the Sun. In Einstein's theory, the Sun warps the fabric of space-time around it. The Earth wants to move along the shortest path because this is what massive bodies naturally do. But in the warped space-time this 'inertial' motion corresponds to an ellipse.

Newton did not show the 'cause' of the apple falling. He showed only that the same force pulls on an apple and the Moon. 'I frame no hypotheses,' he wrote in *The Principia*. Einstein, however, showed the cause of gravity. The Earth warps the space-time around it and the apple and the Moon respond to that warped space-time.

'That one body may act upon another at a distance through a vacuum, without the mediation of anything else, by and through which their action and force may be conveyed from one to another,' said Newton, 'is to me so great an absurdity that I believe no man, who has in philosophical matters a competent faculty of thinking, can ever fall into it.'[15] It was an absurdity. Action at a distance, according to Einstein, was mediated by warped space-time. Newton would have been pleased to be vindicated.

Their visions of space and time provide an even more striking contrast between Newton and Einstein. Newton considered space as a passive backdrop against which the cosmic drama is played out, and time the regular ticking of some universal God-like clock. But, according to Einstein, there is no such thing as absolute space and absolute time. Space and time are stretchy and combined into the seamless entity of space-time. Not only that but matter determines the shape of space-time, which in turn determines how matter moves, which changes the shape of space-time, which changes the way matter moves . . . in the most complex of complex dances. Far from being a passive backdrop to the Universe, space-time is a thing in its own right.

Almost certainly, Newton's view of space and time was pragmatic. He recognised that space could be sensibly defined only as the distance between two bodies. That it must be 'relational'. But he also recognised that, with such a view, it was not possible to make progress with the mathematical tools at his disposal. It was a mark of his genius that he realised that absolute space and absolute time were nevertheless good enough concepts to explain the most obvious features of the Universe.

The voice of space

Space-time's role as an actor in the cosmic drama – as a thing in its own right – gains its most remarkable expression in the phenomenon of 'gravitational waves'. Space-time can be jiggled by the movement of mass. And this jiggling causes waves to propagate outwards like ripples on a pond. Ripples in the very fabric of space-time.

Einstein vacillated about the existence of 'gravitational waves', thinking they existed in 1916, changing his mind shortly after, and then changing it back in 1936. But on 14 September 2015, almost exactly 100 years after Einstein's prediction, gravitational waves were picked up for the first time on Earth.

Imagine being deaf since birth and suddenly, overnight, being able to hear. That is the way it was for astronomers. For all of history they have been able to 'see' the Universe. Now, at last, they can 'hear' it.

Our media has a tendency to overhype things. But a good case can be made that the discovery of gravitational waves is the most important development in astronomy since the invention of the telescope in 1608. They are literally the 'voice of space'.

The event picked up on 14 September 2015 was an extraordinary one. In a galaxy far, far away, at a time when the Earth boasted nothing more complex than a bacterium, two monster black holes were locked in a death spiral. One was 29 times the mass of the Sun and the other 36 times the mass of the Sun. Each travelling at half the speed of light, they whirled about each other one last time. As they kissed and became one, three whole solar masses were destroyed and converted into gravitational waves. A tsunami of tortured space-time surged outwards, so violent that, for a brief instant, its power output was 50 times greater than all the stars in the Universe put together.

Space-time is a billion billion billion times stiffer than steel, which is why it can be vibrated significantly only by the most violent cosmic events such as black hole mergers. But those vibrations, like ripples spreading on a lake, die away rapidly. And the gravitational waves that reached Earth on 14 September 2015, having travelled for 1.3 billion years across space, were mind-bogglingly tiny.

Enter the Laser Interferometric Gravitational wave Observatory (LIGO), in effect a couple of giant 4-kilometre rulers made of laser light – one at Livingston in Louisiana and the other at Hanford in Washington.[16] At 5.51 a.m. Eastern Daylight Time on 14 September 2015, first the Livingston, then 6.9 milliseconds later the Hanford, rulers repeatedly expanded and contracted by 100 millionth the diameter of an atom.[17] 'The signals are infinitesimal. The sources are astronomical. The sensitivities are infinitesimal. The rewards are astronomical,' writes Janna Levin of Columbia University in New York.[18]

The LIGO physicists knew they had detected a burst of

gravitational waves from space because the two detectors, separated by about 2,500 kilometres, picked up precisely the same signal, ruling out the possibility of a mundane local effect like someone slamming a car door 10 kilometres away. But the LIGO physicists were also sure they had detected gravitational waves from space because of the way in which the frequency of the waves rose to a peak then dropped off rapidly as the new-born black hole settled down. It matched precisely the prediction of Einstein's general theory of relativity.

What is so extraordinary about this is that Einstein's theory had previously been tested only in circumstances in which gravity is very weak such as the Solar System, never in the ultra-strong regime that exists in the vicinity of black holes. Yet general relativity passed this test with flying colours. The world's media was quick to declare that Einstein had been proved right. The irony is that he had been proved both right and wrong. Right for predicting gravitational waves. But wrong for not believing in another prediction of his theory of gravity: black holes.

A black hole is surrounded by an imaginary membrane which marks the point of no-return for in-falling light or matter. Just as the sound of a bell ringing is a unique signature of the bell, the sound of this 'event horizon' ringing is the unique signature of a new-born black hole. Because it was heard on 14 September 2015, we now know for sure that black holes exist.[19]

Three men, more than any others, are responsible for LIGO. The first is Kip Thorne of the California Institute of Technology, the hippie-dressing theorist famous for his black hole wagers with Stephen Hawking, most of which he has won. The second is Rainer 'Rai' Weiss, an experimentalist at the Massachusetts Institute of Technology who graduated from building hi-fi sound systems in New York in the 1940s to building sound systems for listening to the cosmos. Weiss has walked every inch of the LIGO tunnels, personally evicting wasps, rats and other intruders. But the most complex and tragic member of the LIGO 'troika' is Scottish physicist Ronald Drever.

A short, dumpy man who carried his papers in supermarket carrier bags and whose overhead projector transparencies were covered with greasy fingerprints and tea stains, Drever was an experimental genius.[20] While Thorne would get an answer to a technical question after pages of careful calculation, Drever would reach the same conclusion with a simple diagram. Unfortunately, the Scottish physicist was constitutionally incapable of sharing control of the project and, in 1997, was fired. He remained in Pasadena, close to Caltech, confused and saddened by events. An unworldly man who never married and had no real friends in the US, he finally succumbed to dementia. In *Black Hole Blues*, Levin relates the heartbreaking tale of Caltech faculty member Peter Goldreich taking the bewildered Drever to New York's JFK airport and putting him on a plane back to his brother in Glasgow. Drever now lives in a care home in Scotland, which means the Nobel committee does not have much time to honour him.

LIGO is a technological marvel. At each site there are actually two tubes 1.2 metres in diameter, which form an L-shape down which a megawatt of laser light travels in a vacuum better than space. At each end the light bounces off 42-kilogram mirrors, suspended by glass fibres just twice the thickness of a human hair and so perfect they reflect 99.999 per cent of the light. It is the Lilliputian movement of these suspended mirrors that signals a passing gravitational wave. So sensitive is the machine that it was knocked off kilter by an earthquake in China. 'It whirs with the tidal pull of the celestial bodies, the grumbling of a still-settling earth, the remnants of heat in the elements, the quantum vibrations and the pressure of the laser,' writes Levin.

A technological marvel LIGO may be, but not everyone thinks it is what it seems. Levin tells of a man on a flight into Baton Rouge, Louisiana, who informed the LIGO scientist in the seat beside him that the secret government facility below them was designed for time travel. 'One of the arms brings you to the future,' he said knowingly, 'the other sends you to the past.'

With the success of LIGO in 2016, we stand at the dawn of a new era in astronomy. It is as if a deaf person has gained a sense of hearing but, at present, that sense is crude and rudimentary. At the very edge of audibility they have heard a distant rumble of thunder. But they are yet to hear birdsong or a piece of music or a baby crying. As LIGO and other gravitational wave experiments around the world ramp up their abilities, who knows what wonders they will soon hear?

Although the announcement of LIGO's direct detection of gravitational waves on 11 February 2016 created huge excitement in the scientific world, compelling indirect evidence of the existence of gravitational waves already existed from the 'binary pulsar' known as PSR B1913+16. In this system, two super-compact 'neutron stars' are spiralling together and so losing orbital energy.

A neutron star is formed in the explosion of a massive star at the end of its life. Paradoxically, while the outer layers of such a star explode into space as a 'supernova', the core of the star implodes, forming a super-dense neutron star relic which typically has the mass of our sun compressed into a volume no bigger than that of Mount Everest. (See page 158 for further information about neutron stars.)

One of the neutron stars in PSR B1913+16 is a 'pulsar' which, as it spins rapidly, sweeps a lighthouse beam of radio waves across the sky. By carefully observing the system, the American astronomers Russell Hulse and Joseph Taylor found that the stars are losing orbital energy at exactly the rate expected if they are radiating gravitational waves. For their discovery, Hulse and Taylor won the 1993 Nobel Prize for Physics.

The mathematics of curved space

In order to turn his basic insight that matter warps space-time, and warped space-time is gravity, into a theory of gravity, Einstein had to wrestle with the complex mathematics of curved space. Unfortunately, he had skipped mathematics lectures while a student at the Swiss Federal Polytechnic in Zurich, preferring instead to get his hands dirty among the batteries and condensers and galvanometers of the Polytechnic's Electrotechnical Laboratory. 'It was a mistake I realised only later, with regret,' said Einstein.[21]

Fortunately, Einstein had made a lifelong friend in Marcel Grossman, a mathematics student a year his senior at the Polytechnic. Grossman's father had used his contacts to help secure Einstein his dream job at the Swiss Federal Patent Office in Bern. And Grossman, crucially, knew about the geometry of curved spaces. He was therefore able to teach Einstein the mathematics he needed in order to express in rigorous terms his revolutionary ideas about gravity and the warpage of space-time.

This field of mathematics had been developed by a number of mathematicians, most importantly Carl Friedrich Gauss and Bernhard Riemann in the nineteenth century. Until that time, geometry was the flat-paper geometry of the Greek mathematician Euclid (who inspired the best-ever title of a popular-science book, *Here's Looking at Euclid!*).[22] In his *Elements*, written in the third century BC, Euclid listed five self-evident statements about straight lines and angles. Using these 'axioms' as foundations, using logic alone he constructed a great edifice of 'theorems' such as 'the internal angles of a triangle always add up to 180 degrees'.

Euclid's fifth postulate states that parallel lines never meet. Gauss and Riemann relaxed this postulate, which immediately admitted geometries of curved surfaces such as spheres. On a sphere, for instance, two parallel lines which extend northward from the equator do not stay parallel but meet at the North Pole.

Einstein in Berlin

The struggle to describe gravity as warped space-time – or, more generally, to generalise relativity – would take Einstein eight long years. During that time, he moved from Zurich to Berlin.

Einstein had actually been born in Germany in the southern city of Ulm. But his dismay at the country's militarism had caused him to renounce his German citizenship in 1896, aged twenty. Despite this, he accepted a university post in Berlin, a city he would call home from 1914 until Hitler's seizure of power in 1933 made it impossible for Jews like him to remain without endangering their lives, and he emigrated to the United States.

Einstein was lured to Berlin by Max Planck and Walther Nernst. The two giants of German and world science had arrived at the train station in Zurich with an offer Einstein could not refuse: a lucrative professorship at the University of Berlin with no teaching responsibilities. Berlin was fast becoming the epicentre of the scientific world and the possibility of conversing on a daily basis with some of the best scientists on the planet proved a mouth-watering prospect to a man who had spent years in the intellectual wilderness of the Swiss Federal Patent Office. Berlin also offered the possibility of breaking free from the shackles of his by now broken and dysfunctional marriage.

Einstein's ascent into the scientific stratosphere had been paralleled by Mileva's descent into a world of childcare and household drudgery. If this was not enough to drive a wedge between husband and wife, Einstein proved himself fundamentally unsuited to marriage. He found it impossible to apply the degree of concentration needed to make profound scientific discoveries and simultaneously occupy his mind with anything else, be it the trivia of everyday life or a meaningful personal relationship.

Newton had solved this problem by never marrying and, as far as anyone knows, never having any close personal relationships. Einstein, though priding himself on his unconventionality, had succumbed to convention, and out of duty married Mileva, though some time after she had become pregnant and given birth to a baby that had been spirited away to Serbia and Mileva's family. Grief for the loss of a child, whose existence was largely kept a secret from friends, must have put significant strain on the marriage. But the truth is that it had not been the happy and fulfilling union between equals that the two had imagined while naive students at the Swiss Federal Polytechnic.

Einstein travelled from Zurich to Berlin on a circuitous route that involved him visiting physicist friends around Europe. At last he arrived in the Prussian capital in April 1914 and his family soon joined him there. But, by early July, things had gone badly wrong between him and Mileva, and she returned to Zurich with the children. Though the Einsteins did not divorce until 1919, their marriage was effectively over.

In Berlin, Einstein rekindled an affair with his cousin Elsa, which had begun and ended a few years earlier. A divorcee with few prospects, Elsa cooked and cleaned for Einstein and was willing to accept what Mileva could not: that in exchange for the prestige of being with such a renowned man, she was otherwise to place no demands on him or his time.

Einstein treated Mileva abominably. Nevertheless, he was tearful as his wife and two sons boarded the train to return to Zurich. But back at his empty apartment in the suburb of Dahlem, he sat down at his desk and began to work. He had achieved what he most wanted in the world: a bachelor life free of domestic distraction and family responsibility. His friend Janos Plesch described it this way: 'He sleeps until he is awakened; he stays awake until he is told to go to bed; he will go hungry until he is given something to eat; and then he eats until he is stopped.'

Einstein believed he was at last at peace. He was sadly mistaken.

Within weeks, Germany and its allies were at war with Russia, the British Empire and France. Einstein was in a state of shock. And his shock was magnified by the overnight transformation of his fellow physicists into a war-drunk mob. 'Our whole, highly prized technological progress and civilisation can be likened to an axe in the hand of a pathological criminal,' said Einstein.[23]

Most distressing of all was the behaviour of the chemist Fritz Haber, his closest friend in Berlin. Haber had tried to be a marriage guidance counsellor for Einstein and Mileva, and had accompanied Einstein and his departing family to the train station in Berlin. Now Haber turned his laboratory into a military factory and began concocting ever more horrible poison gases with which to dispense agony to the young men of Europe.[24]

In the midst of a catastrophic war, Einstein's monumental detachment – which had wrecked his marriage – served him incredibly well. Locked away in his office in Haber's institute, surrounded by chemists-turned-killers, he lost himself in the world of physics and, in particular, his theory of gravity.

Einstein delivered his first lectures on the new theory to the Prussian Academy in October 1914. It was not yet complete. But Einstein was confident enough not only to declare that the great Isaac Newton was wrong but that the geometry of warped space-time was crucial for understanding gravity. He might as well have been talking in Martian. Despite being a supernova in the firmament of physics, nobody in his audience took much notice of him. Einstein, being Einstein, was utterly unperturbed. He went back to his office, shut the door, and went back to work.

A year later, at the end of 1915, everything came to a head.

November 1915: Hilbert

Einstein had been invited to give a series of lectures at the University of Göttingen by the foremost German mathematician of his day. David Hilbert had achieved enduring fame in 1900 by high-lighting what he considered to be the twenty-three outstanding problems in the field – setting out a road map for mathematical research in the twentieth century.

Ignored by his colleagues in Berlin, Einstein jumped at the chance of being listened to in Göttingen. He travelled to the university town and, at the end of June and the beginning of July 1915, delivered six lectures on his theory of gravity. He informed his audience that the transformation of gravity into geometry was largely correct – though it was still not 100 per cent right. In particular, his theory of gravity was incompatible with one of the key elements of his special theory of relativity of 1905: that observers in uniform motion relative to each other should see the same laws of physics. It also had the problem that it did not predict the correct orbit for Mercury.

Hilbert was sure Einstein was on the right track, so Einstein was in high spirits when he returned to Berlin. But, by late September, his joy had turned to horror.

Unusually for a mathematician, Hilbert was also deeply interested in physics. It was the reason he had invited Einstein to Göttingen in the first place. And it was his interest in physics that prompted him to try and fix the problems Einstein had highlighted in his lectures. Dropping what he was doing, he began looking for a theory of gravity that was compatible with the special theory of relativity. After eight years working entirely alone,

Einstein had a competitor. And not any competitor but a man of exceptional mathematical ability.

Still worse, by the end of September, it dawned on Einstein that the incompatibility of his theory with special relativity and its inability to predict the orbit of Mercury were not mere details, as he had naively imagined. They were fundamental. In particular, observers rotating relative to each other would see different laws of physics, which was incorrect. His theory of gravity was in deep trouble.

Einstein, understandably, was depressed. And he could easily have succumbed to the pressure. But, his depression quickly turned to rage. There was no way in the world that he was going to let someone else take the credit for his eight long years of toil. Not without a fight.

By the first week of October, a miracle had occurred. Einstein saw how to proceed. 'A good scientist is someone who works hard enough to make every possible mistake before coming to the right answer,' said the American physicist Richard Feynman.[25] Einstein was that scientist. He made *every possible mistake* in his long struggle to obtain his theory of gravity. But the mark of his genius was that every time he found himself hopelessly lost in the forest of the night, he somehow found a path out again.

Out of the woods and back on the right trail, Einstein worked in a frenzy for the next six weeks, often forgetting to eat or sleep. It was, he would later maintain, the most intense mental struggle of his life.

By the beginning of November, he was almost, but not quite, there. He still did not have the correct equation for describing the gravitational 'field'. But he could not afford to delay going public a moment longer.

Einstein had agreed months before to present his theory in a series of lectures at the Prussian Academy. At the time, he had thought his theory was in good shape. Now, of course, he knew that it was incomplete. Nevertheless, he had to go ahead. It was a race against time. He simply had to get to the finishing post before Hilbert.

Einstein was to a give one lecture a week for four weeks. He managed to get together enough material to give the first lecture. From then on, it was seat-of-the-pants stuff. During each succeeding week, he spent his time furiously trying to solve the problem that he had struggled with for eight years. At the end of each week, he stood before his audience at the Prussian Academy and lectured on what he had just figured out.

All the while, his rival was breathing down his neck. The letters from Hilbert showed that the great mathematician was more or less on the right track. They spurred Einstein to ever more frantic activity.

In his first lecture, delivered on 4 November, Einstein made no prediction. But now the theory was internally consistent and compatible with the special theory of relativity. As if to underline the fact, Einstein was able to show that Newton's theory of gravity emerged as an approximation of his theory when the curvature of space-time is small.[26] For the first time the theory had the smell of something that was right.

Two weeks later, on 18 November 1915, Einstein at last announced some predictions of his theory. He had calculated the gravitational field close to the Sun. This enabled him not only to calculate the light bending by the Sun but, most importantly, to predict the precession of the perihelion of Mercury.

The anomalous motion of Mercury

On Christmas Eve 1907, just after he had completed his review of special relativity, Einstein had written to his Zurich friend Conrad Habicht: 'I hope to explain the still unexplained secular changes in the perihelion distance of Mercury.'[27] Back then, he had failed. Nevertheless, the letter reveals Einstein's prescience in recognising that such a tiny effect was a subtle symptom of a fundamental failure of Newton's theory of gravity.

Mercury is the Sun's closest companion. Being so near the most massive body of all, the planet is forced to negotiate the most grossly warped space-time in the Solar System. It makes it the planet on which the effects of warped space-time leave their biggest mark.

In 1905, Einstein had discovered that all forms of energy have an effective mass. It follows that all forms of energy must exert gravity. And one of those forms of energy is gravitational energy – the energy of warped space-time itself. Remarkably, this means that warped space-time is not only gravity but a *source of more gravity*. Gravity creates more gravity!

Close to the Sun, therefore, gravity is stronger than predicted by Newton. It departs from a force described by an inverse-square law.

Newton's great triumph was of course his demonstration that a body experiencing a centrally directed inverse-square law of force travels in an ellipse. It obviously follows that, if a body does not experience an inverse-square law of force, it does not travel in an ellipse. The trajectory is instead an ellipse that 'precesses', continually changing its orientation in space, and tracing out a rosette-like pattern.

Einstein calculated Mercury's orbit. His theory predicted that the effect of warped space-time near the Sun should indeed cause an anomalous precession of Mercury's orbit. The amount was 43 arc seconds per century.

It was exactly the anomalous precession that had mystified astronomers for half a century. It was exactly the precession that had prompted Urbain Le Verrier to postulate the existence of the planet Vulcan.

There was, of course, no Vulcan. The anomalous motion of Mercury was not telling astronomers of the existence of an unknown planet skirting the fires of the Sun. It was signalling something far more fundamental and astonishing – something that nobody before had ever suspected: that Isaac Newton was wrong.

'The theory agrees completely with the observations,' Einstein concluded on presenting his Mercury result to the Prussian Academy. He had overturned 200 years of physics and shown that the greatest scientist ever to have lived had been wrong but he had managed to disguise what he was actually feeling. Inside he was in utter turmoil. He was beside himself with excitement.[28] In fact, he was having palpitations.[29]

Physicists scrawl arcane mathematical equations across black-boards but it is an enormous leap of faith to believe that nature really obeys those equations. It invariably comes as a massive shock when it turns out that nature really does.

After an eight-year struggle, Einstein had finally reached the summit of a towering mountain. The fog that had enshrouded him every step of the climb had cleared. And stretching out below him, lit by dazzling sunlight, was a landscape no other human being had ever seen. 'The years of searching in the dark for the truth that one feels but cannot express,' said Einstein, 'the intense desire and alternations of confidence and misgiving until one breaks through to clarity are known only to him who has experienced them.'[30]

Actually, Einstein was not the only one to suggest that the anomalous motion of Mercury could be explained if gravity close to the Sun was slightly greater than predicted by Newton's law of gravity. In the late nineteenth century, the American astronomer Simon Newcomb had pointed out that the planet's motion made sense if gravity weakened not according to the inverse-square of the distance between masses – that is, according to a power of 2 – but according to a power of 2.0000001612.[31, 32]

Substituting 2.0000001612 for 2 marred the simplicity of Newton's law of gravity. But, if nature chose ugly over beautiful, there was no choice but to accept it. What scuppered Newcomb's idea was the recognition that, although such a messy 'power law' of gravity might be able to explain the motion of Mercury, it could do it only at the expense of not being able to explain the motion of the Moon.

Einstein's explanation predicted the observations of both Mercury and the Moon. Close to the enormous mass of the Sun, space-time was warped enough to cause a noticeable anomaly in the motion of the innermost planet. Close to the far punier Earth, space-time was far less warped, so there was no noticeable anomaly in the motion of the Moon.

It was history repeating itself. Hendrik Lorentz and George FitzGerald had proposed that the length of a body contracts at speeds approaching the speed of light but had provided no fundamental explanation for it. Einstein did. Here, Newcomb had proposed that gravity close to the Sun was slightly stronger than Newton predicted but provided no fundamental – or, in this case, correct – explanation for it. Einstein did.

Einstein's field equations

The pressure from Hilbert breathing down Einstein's neck had had the desired effect. In the week before his fourth and final lecture, after eight years of struggle, and in the absolute nick of time, Einstein reached his goal. On 25 November 1915, his coat buttoned up against the cold, he made his way along Unter den Linden Strasse and addressed his audience at the Prussian Academy. On the blackboard he simply wrote:

$$G_{uv} = 8\pi G T_{uv}/c^4$$

It is the law of gravity everyone experiences, no matter what their state of motion. It is the general theory of relativity in a nutshell. American science writer Dennis Overbye describes it as 'the equation that rules the Universe'.[33]

Einstein's equation uses super-compacted notation so, like Dr Who's 'Tardis', it is bigger on the inside than on the outside. The left-hand side of the equation is in fact a 4 × 4 table of numbers known as the 'Einstein curvature tensor', which summarises the curvature of space-time. The right-hand side is another 4 × 4 table of numbers known as the 'stress-energy tensor', which summarises the 'sources of gravity'.[34]

That Einstein's equations contain 4 × 4 tables of numbers means that there are actually sixteen equations. In fact, Einstein was able to use 'symmetry arguments' to reduce the number of equations down to just ten. Nevertheless, the fact remains that he still substituted ten equations for the single equation of Newton's theory of gravity.

Einstein's 'field equations of gravity' dictate the warped space-times that are generated by any distribution of mass-energy. They are the mathematical embodiment of John Wheeler's phrase:

'Matter tells space-time how to warp. And warped space-time tells matter how to move.' Finding space-times that satisfy the ten gravitational field equations is extremely hard. In fact, it is so hard that anyone who finds one often has the space-time named after them.

Einstein's field equations are 'generally covariant', which means they do not depend on your point of view (technically, they retain their form no matter what system of coordinates they are expressed in). This is their beauty. And this is something Einstein had shed blood and tears to achieve.

But Einstein's theory was not quite the one he originally set out to find in 1907. His aim had been to generalise his special theory of relativity by finding what must be done to the measurements of space and time of people varying their speed, or 'accelerating', relative to each other so that they would agree on the same laws of physics. In fact, Einstein replaced Newton's law of gravity with a new and improved theory of gravity rather than finding a theory about accelerated observers. Such is the serendipity of science.

The bending of light by gravity

The scene, as Einstein's chalk squeaked across a blackboard in Berlin, could not have contrasted more starkly with the outside world, where the slaughter of young men on an industrial scale had been gaining momentum. Already in 1915, gas attacks had poisoned, burnt and suffocated soldiers on all sides; Zeppelins had rained down death on British civilians; and a U-boat had torpedoed the ocean liner *Lusitania* off the coast of Ireland, with the loss of 1,198 lives.

But despite the mounting horrors, contact, incredibly, was maintained between the scientists of the warring nations. Within weeks of the publication of the general theory of relativity, copies were smuggled out of Germany to Holland, and then to England. And though the 'war to end all wars' would leave 10 million dead and as many again with their health wrecked for ever, within a year of the Armistice on 11 November 1918 it was an Englishman who confirmed a key prediction of Einstein's, catapulting the German-born physicist into the scientific firmament.[35]

Arthur Stanley Eddington had received his smuggled-out copy of Einstein's theory from the Dutch astronomer Willem de Sitter in Leiden. An accomplished populariser of science, the Cambridge scientist became the chief communicator of Einstein's ideas to the English-speaking world. Asked by a journalist in 1919: 'Is it true that only three people in the world understand the theory of general relativity?' he replied (perhaps not as modestly as he might have done): 'Oh. And who is the third?'

Eddington zeroed in on Einstein's prediction of light bending by the gravity of the Sun. Einstein had realised the effect in 1907 when he had written his review article on the special theory of relativity, and had begun to think for the first time about developing a theory of gravity that, unlike Newton's, was compatible with the new view of space, time, matter and energy.

The special theory of relativity had revealed that all energy – including light energy – has an effective mass.[36] Consequently, a massive body like the Sun must attract light as surely as it attracts matter. Observing this effect would provide strong evidence for Einstein's theory of gravity.

However, by the time Einstein had developed his full-blown theory of gravity, he realised that the bending of light by gravity was actually a more subtle effect than he had guessed in 1907.

Let's return to the astronaut in the blacked-out cabin of a rocket accelerating at 1g, far away from the gravity of any planet. Because the astronaut's feet are pinned to the floor and all objects fall at the same rate, irrespective of their mass, there is no way he can tell that he is not on the surface of the Earth.

Well, actually, that is not entirely true. There is one way he can tell.

The Earth is round. Consequently, all bodies fall towards the centre of the Earth. In the most extreme case, when objects are dropped on opposite sides of the globe – say, in England and New Zealand – they fall in opposite directions. Actually, wherever two objects are dropped their paths inevitably converge as they head towards the centre of the Earth.

But this is not what the astronaut in the rocket sees. If he observes two falling objects with a precise enough measuring device, he finds that their paths do not converge but stay parallel. So he is able to guess that he is not on the surface of the Earth.

Remarkably, this is not a killer blow for Einstein's theory of gravity. The Principle of Equivalence on which the great edifice of the general theory of relativity is built actually requires only that gravity and acceleration are indistinguishable *locally* – that is, in an arbitrarily small region of space.

But the fact that objects fall along converging paths in the vicinity of a real body such as the Earth or Sun has implications for the path of a light beam. In the neighbourhood of such bodies – as opposed to inside the cabin of the astronaut's rocket – the beam bends by twice as much as naively expected.

The body in our vicinity with the greatest light-bending power is of course the Sun, which contains 99.8 per cent of the mass of the Solar System. The best way to see the effect, Einstein realised, is to observe a distant star whose light, on its way to the Earth, passes close to the solar disc, where the valley of space-time is at its steepest. The path of the light will be bent as surely as the path of a hiker negotiating a hilly landscape. If the star is observed from Earth, it will be shifted from its expected position in the sky.

A tale of two eclipses

Obviously, stars close to the Sun are lost in its glare, as impossible to see as fireflies next to a car headlight. But there is one instance in which such stars can be seen: when the luminous disc of the Sun is blotted out by the disc of the Moon. In such a 'total eclipse', the world is plunged into darkness and for a few minutes, the stars come out in daytime.

A total eclipse occurs somewhere in the world every few years. But the necessary alignment between the Sun, the Moon and the

Earth occurs only along a narrow 'track' on the Earth's surface. Consequently, the chance of seeing a total eclipse at a given spot in a given year is very small – there is one only every 350 years on average.

As luck would have it, a total eclipse was visible from Russia's Crimea peninsula – not too far from Germany – on 24 August 1914. A German expedition was mounted, led by Erwin Freundlich, an astronomer who was greatly impressed by Einstein's ideas. On 19 July, Freundlich left Berlin with two companions and four telescopes equipped with cameras. It was a bad time to be going to Russia.

Freundlich had probably heard of the shooting three weeks earlier in Sarajevo of Austrian Archduke Franz Ferdinand by a Serbian nationalist. But, in common with everyone else in Europe, he had no inkling of the disastrous chain of events Gavrilo Princip had set in motion. On 1 August, three days before Britain did the same, Russia declared war on Germany.

Overnight, Freundlich and his companions were transformed from guests of the Russians into enemy aliens. Their equipment was impounded and they were imprisoned. Consequently, they missed the total eclipse – which, anyhow, was obscured by clouds over the Crimean peninsula. But their misery was short-lived. In one of the first prisoner exchanges of the First World War, they were swapped for Russian officers, and they limped back to Berlin by the end of September.

For Einstein this was actually a piece of good fortune – and not only because Freundlich was a friend and supporter. Had the astronomer succeeded in measuring the deflection of starlight by the Sun, it would not have matched Einstein's prediction. The reason was that, in 1914, Einstein still believed that the deflection would be 0.87 arc seconds – the figure he had obtained in 1911 – rather than the correct figure of 1.7 arc seconds he obtained from his full theory in 1915.[37]

The First World War finished and, on 29 May 1919, there was another total eclipse of the Sun. Eddington and an assistant set off for Príncipe, a small volcanic island in the Gulf of Guinea off the coast of West Africa. Weather conditions were not brilliant on 29 May. In fact, the morning began with a tropical downpour. But, although the rain eased off by the time of the eclipse in the early afternoon, Eddington and his assistant were dismayed to see the clouds thicken and clear repeatedly as the Sun was blotted out by the black disc of the Moon. There was nothing for it but to carry on regardless taking photographs and hope for the best.

Of the sixteen exposures obtained by Eddington only six were made in cloud-free conditions. Four of them could not be developed in the tropical heat of Príncipe but had to be packed away for transport back to England. Of the remaining two exposures, only one had captured a clear enough star-filled sky for Eddington to carry out the necessary measurements.

But one was all he needed.

On 3 June, Eddington compared the star positions recorded during the total eclipse with their positions recorded on a photograph taken back at Greenwich in England. It was a difficult measurement to make. A single arc second on the sky corresponded to a mere 1/16 millimetre on Eddington's photographic plate. But the English astronomer rose to the challenge. He made his painstaking measurement. He checked it and re-checked it.

There was no doubt about it. The stars near the Sun were shifted in position by 1.61 arc seconds, plus or minus 0.3 arc seconds. It was within a whisker of Einstein's prediction.

Eddington would look back on this magical moment as the single most important incident of his life. He had confirmed the general theory of relativity. Newton was wrong. A forty-year-old German had assumed his mantle. Eddington penned the following ditty:

> One thing at least is certain, light has weight
> Light rays, when near the Sun, do not go straight.

Bizarrely, an eclipse expedition in 1914 which had failed because of a man called Princip had been followed by one in 1919 which succeeded on the island of Príncipe.

Einstein was sick and in bed when a telegram reached him

from his friend Hendrik Lorentz. It did not quite say that general relativity had been confirmed. But it probably relayed the charming but brief wording of the telegram Eddington had sent from Príncipe back to England:

Through cloud. Hopeful.
Eddington.

It was enough. 'I knew I was right!' exclaimed Einstein.[38]

And Einstein did know he was right. It was not simply that he was cocksure – though he was certianly that – but he had a great belief that nature's fundamental laws must be elegant and beautiful. And the equations of general relativity were certainly that. Later, he was asked by a doctoral student: 'What if general relativity had not been confirmed by Eddington?'

'Then I would feel sorry for the dear Lord,' replied Einstein.[39]

On 7 November 1919, on page 12 of *The Times* of London, there appeared an article under a triple-headline:

REVOLUTION IN SCIENCE

New Theory of the Universe

Newtonian Ideas Overthrown

It was the report of a joint meeting of the Royal Society and the Royal Astronomical Society which had been held the day before. Overnight, Einstein became a superstar. He was destined to achieve the global fame of Charlie Chaplin. In fact, he would stay with Chaplin and his wife when visiting Los Angeles.[40] Such would be his fame that when Edith Piaf made her first visit to America in 1947 and was asked at a press conference who she would most like to meet she did not hesitate: 'Einstein. And I'm counting on you to get me his phone number.'[41]

On Einstein's first visit to London in 1921, he stayed in the home of the biologist J. B. S. Haldane. It was like having The Beatles in their screaming heyday come to stay. So overcome with excitement and hysteria was Haldane's daughter at seeing Einstein walk through the front door that she promptly fainted.[42]

The morning before the lecture that he had come to London to deliver, Einstein left Haldane's house and walked to Westminster Abbey. There, in the nave against the choir screen, he gazed on the marble tomb of his great predecessor: Isaac Newton.

Both Newton and Einstein had been inspired by falling bodies to create their theories of gravity. In the fall of an apple Newton had seen the fall of the Moon, and unified the earth with the heavens. In the fall of a man from a roof, Einstein had seen that the force of gravity is an illusion. Both men had known what it was like to 'voyage through strange seas of thought alone'. 'Nature to him was an open book, whose letters he could read without effort,' said Einstein. What would he have given to have met Newton – a man who had died two centuries earlier but whose thought processes he understood better than any man alive?

With the general theory of relativity, Einstein now had in his hands one of the most powerful tools in the history of physics. But though he was a genius, he was not infallible. And, remarkably, he would miss some of the most important predictions of his own theory. Those predictions – black holes and the big bang – would reveal that Einstein's theory of gravity, though a huge improvement on Newton's, is itself flawed.

Further reading

Einstein, Albert, *Relativity: The Special and General Theory*, Folio Society, London, 2004.

Fölsing, Albrecht, *Albert Einstein*, Penguin, London, 1998.

Levenson, Thomas, *Einstein in Berlin*, Bantam Books, New York, 2003.

Levenson, Thomas, *The Hunt for Vulcan . . . And how Albert Einstein destroyed a planet, discovered relativity and deciphered the Universe*, Head of Zeus, London, 2015.

Levin, Janna, *Black Hole Blues*, The Bodley Head, London, 2016.

Overbye, Dennis, *Einstein in Love: A Scientific Romance*, Viking, London, 2000.
Pais, Abraham, *'Subtle is the Lord . . .': The Science and the Life of Albert Einstein*, Oxford University Press, Oxford, 1983.

ʎ

Where God divided by zero

*How Einstein's theory of gravity predicts daft
things at the 'singularity' of a black hole and
how a deeper theory is needed that doesn't*

For years, my early work with Roger Penrose seemed to
be a disaster for science. It showed that the Universe must
have begun with a singularity, if Einstein's general theory of
relativity is correct. That appeared to indicate that science
could not predict how the Universe would begin.

Stephen Hawking[1]

The black hole teaches us that space can be crumpled like
a piece of paper into an infinitesimal dot, that time can be
extinguished like a blown-out flame, and that the laws of
physics that we regard as 'sacred,' as immutable, are anything but.

John Wheeler[2]

In February 1916, Einstein received a surprising package. It came
from a soldier serving on the Eastern Front. Karl Schwarzschild
had been the director of the Astrophysical Observatory in Pots-
dam, just outside Berlin. But at the outbreak of war in 1914 he
had been overcome by extreme patriotic fervour and dropped
everything to volunteer for military service. In his eighteen
months in the Kaiser's army, he had run a weather station in
Belgium, calculated shell trajectories with an artillery battery in
France, and now he was serving in Russia.

Despite being caught up in a vicious war, Schwarzschild

found time to write several scientific papers, two of which were on Einstein's theory of gravity, which he had learnt about soon after its publication at the end of 1915. What was notable about Schwarzschild's work was that in such a short time he had taken a significant step beyond Einstein.

The equations of general relativity are complex. They substitute a total of ten equations for Newton's single inverse-square law formula. Because of their complexity it is very hard to deduce the shape of the space-time in the vicinity of a realistic body. But Schwarzschild made a number of simplifying assumptions, which reduced Einstein's equations to a more straightforward and manageable form, and enabled him to 'solve them'.

Schwarzschild's 'solution' described the shape of the warped space-time in the neighbourhood of a localised mass such as a star. Einstein was amazed. 'I had not expected that one could formulate the exact solution of the problem in such a simple way,' he wrote back to Schwarzschild.

Most remarkably, Schwarzschild showed that if enough mass were crammed into a small enough volume, space-time would become so extraordinarily warped as to become a bottomless well. So steep would be the sides of the well that a light beam trying to climb out would die of exhaustion, sapped of all its energy, before it could escape. With no light emerging, the region of space-time would appear blacker than night.

Schwarzschild had no word for what he had discovered. That would only be coined by the American physicist John Wheeler in 1967. But today there is hardly a person alive who does not know the term. Schwarzschild's solution described a 'black hole'.[3, 4]

Schwarzschild's story is a tragic one. While in Russia, he developed a rare and serious 'autoimmune' disease in which the body's immune system malfunctions and attacks healthy tissue. *Pemphigus vulgaris* causes painful blisters on the skin as well as inside the mouth, nose, throat, anus and genitals. No one knows its cause – although it may be a combination of genetic and environmental factors – and there is no cure – although modern treatments with corticosteroids relieve the symptoms. If the blisters become infected, the infection can spread into the bloodstream and affect the whole body. This is what happened to Schwarzschild. He was invalided back to Berlin in March 1916 but died two months later on 11 May. He was forty-two.

Schwarzschild's black hole was surrounded by an 'event horizon'. Anything passing through this into the interior – whether light or matter – could never get out again. The event horizon provides a measure of the 'size' of a black hole. For the Sun to become a black hole, it would have to be crushed within a sphere only 3 kilometres in radius. For the Earth, the 'Schwarzschild radius' would be a mere 2 centimetres. Fortunately for us, the Sun and the Earth are not massive enough for their gravity ever to turn them into black holes.

But if a very massive star was crushed within its event horizon – causing it to wink out of existence as far as the rest of the Universe was concerned – its gravity would continue to crush the star all the way down to an infinitesimal point. With the star effectively snuffed out, all that would be left behind would be a bottomless well of space-time. 'The black holes of nature are the most perfect macroscopic objects there are in the Universe,' said the Indian Nobel prizewinner Subrahmanyan Chandrasekhar. 'The only elements in their construction are our concepts of space and time.'[5]

At the centre of the black hole, where the matter of a star is crushed to infinite density, the curvature of space-time and the strength of gravity skyrocket to infinity.[6] 'Black holes are where God divided by zero,' observed American actor and writer Stephen Wright. The appearance of such a nonsensical 'singularity' in any theory indicates that it no longer describes reality. It has broken down, become indistinguishable from gobbledygook.

Of Schwarzschild's black hole paper, Einstein said: 'If this result were real, it would be a true disaster.' But not for an instant did Einstein – or even Schwarzschild – think the result was real. Neither did they entertain the thought that the black hole solution described an object that might actually exist in the Universe.

The few who did were not overly worried either. A star has a finite supply of energy and, when it exhausts it, its internal fires must go out. Since those fires push outwards and prevent gravity from crushing the star during its lifetime, the star will begin shrinking down to a singularity. But some new force was bound to come to the rescue and halt the shrinkage long before that happened. It was inconceivable that nature would permit the formation of a monstrous singularity.

Actually, it appeared nature did indeed provide such a force. It was a consequence of 'quantum theory', the bizarre theory of the microscopic world of atoms and their constituents.[7]

Quantum stars

Quantum theory was stumbled on in the first decades of the twentieth century but given a firm mathematical foundation only in the mid-1920s. The theory recognised that the fundamental building blocks of matter behave both as localised particles – like tiny billiard balls – and spread-out waves – like ripples on a pond. This peculiar 'wave–particle duality' leads to a multitude of strange and unexpected phenomena – for instance, the ability of a single particle to be in two or more places at once. It also plays a crucial role when a star at the end of its life runs out of the fuel necessary to maintain its internal fires.[8]

Robbed of its ability to push outwards, the matter of a star is crushed by the iron fist of gravity until it fills a volume of about the size of the Earth. Such a 'white dwarf', 100 times smaller and about a million times denser than the Sun, is the endpoint of the evolution of all normal stars, including our own. In such super-dense conditions – a sugar-cube-sized volume of white dwarf stuff weighs as much as a family car. – the electrons are forced very close together.

Squeezing a wave of any type into a small space causes it to become more choppy and violent. In the case of a quantum wave, more choppy and violent corresponds to a faster-moving particle (strictly speaking, one with greater 'momentum'). This is the famous 'Heisenberg Uncertainty Principle'. And it dictates that when electrons are squeezed tightly together inside a white dwarf, they attain extremely high speeds.

This is one quantum effect with important implications for

white dwarfs. But there is a second one, which is a bit harder to explain.[9] Take it on trust that another consequence of wave–particle duality is that the fundamental building blocks of matter come in two distinct tribes: 'bosons', which are gregarious; and 'fermions', which are antisocial. Fermions, which include the electron, are said to obey the 'Pauli Exclusion Principle', which states that no two fermions can occupy the same quantum 'state'.[10]

For electrons in a white dwarf this means that two neighbouring particles must have distinctly different velocities. So, if one has a velocity dictated by the Heisenberg Uncertainty Principle, its neighbour must have an even higher velocity – in practice, double the velocity; its neighbour an even higher velocity – in practice, triple the velocity; and so on.

Picture a ladder, with each rung corresponding to a higher and higher velocity. According to the Pauli Exclusion Principle, there can be only one electron on each rung (actually, there can be two, but that is another story!).[11] The Principle ensures that the electrons in a white dwarf have extraordinarily high velocities, boosted way beyond what the Heisenberg Uncertainty Principle might suggest. And it is these super-fast electrons buzzing about inside the stars that push back against gravity. Their so-called 'electron degeneracy' pressure keeps a white dwarf stable and prevents it from shrinking to a ball much smaller than the Earth.[12]

This was the state of play in the late-1920s. Quantum theory, miraculously, came to the rescue of a dying star. It staved off the runaway collapse down to a black hole with a nightmarish singularity in its heart. All was under control. Everything in the garden was rosy.

Or so it seemed.

The Chandrasekhar limit

In August 1930, a nineteen-year-old Indian embarked on a ship in Bombay bound for England and the University of Cambridge. I have already quoted his older self on the subject of the remarkable simplicity of black holes. His name was Subrahmanyan

Chandrasekhar and he was something of a mathematical prodigy.

The voyage was initially assailed by bad weather and the ship had to steam at half-speed. But at Aden the sun came out. And as the ship made its way through the Suez Canal, Chandrasekhar was at last able to leave his cabin, where he had been largely imprisoned during the heavy seas.

I imagine him cutting an eccentric figure as he staggers out on deck carrying a teetering pile of books on quantum theory and astrophysics. Perspiring copiously, he dumps the books in one deckchair and collapses in another beside it. Other Indians promenading past shoot him odd looks. It is no more than he expects. He has made no effort to interact with any of his fellow countrymen and he is acutely aware that they consider him aloof, if not arrogant. But he does not care. At last, he has the peace and quiet to think, really think. And what he thinks about, incongruously, as the sands of the Sinai Peninsula sail past and the hot desert air scours his face, is white dwarfs. One question and one question alone occupied Chandrasekhar's mind: were the electrons in a white dwarf relativistic? Flicking back and forth between his books and papers, he gathered together the formulae that described the interiors of stars and the quantum behaviour of electrons at ultra-high density. He put in the numbers he knew and cranked away until finally there emerged an answer. He checked and checked again. There was absolutely no doubt about it. The electrons inside a white dwarf would be moving at more than half the speed of light, a velocity at which the effects of Einstein's special theory of relativity would be apparent. In the jargon, they were 'relativistic'.

Such velocities were staggeringly huge: more than 150,000 kilometres per second. But more important to Chandrasekhar was the implication of such speeds. It meant that quantum theory alone was insufficient to understand white dwarfs. A correct theory must also incorporate Einstein's special theory of relativity.

When night fell the sky was crowded with an impossible number of stars. Nobody guessed that the strange young man in the deckchair, so engrossed in his notebooks he often forgot to go to dinner, was calculating the properties of their interiors. His body may have been pinned to the deck of a ship but his mind ranged freely among the embers of dying suns.

It did not take long for Chandrasekhar to develop a properly relativistic theory of white dwarfs. And it did not take long for him to discover something unexpected and extraordinary, if not downright horrifying.

The more massive a white dwarf, the more its gravity squeezed the electrons in its interior and the faster they buzzed about. That much was true. Except that Einstein's theory of relativity imposed a limit on how fast the electrons could go: the speed of light. As the electrons approached the cosmic speed limit, they became ever more massive and it became more and more difficult to boost their speed. But this created a problem. After all, it was the continual drumming of the electrons – like raindrops on a tin roof – that provided the outward force to oppose the gravity trying to crush a star. If, as they were squeezed harder and harder, their speeds were boosted by ever smaller amounts, their ability to oppose gravity was gradually drained away. The young Indian in the deckchair, with his head in the stars, saw the looming stellar catastrophe like a train bearing down on him in the night.

For a white dwarf, the stiffness of the electron gas holding back gravity was like the stiffness of a cricket ball resisting a bowler's grip. But, above a certain stellar mass, everything changed. The cricket ball abruptly turned into marshmallow.

Chandrasekhar did the calculation, over and over again, checking and rechecking that he had made no mistake. But there was no doubt about it. If a star at the end of its life were more massive than 1.4 times the mass of the Sun, electron degeneracy pressure would not be enough to save it. Gravity would crush the star catastrophically. No known force in the Universe could stop it. The monstrous singularity was unavoidable.

Neutron stars

Two years later, in 1932, the English physicist James Chadwick found a particle as massive as the positively charged proton, but with no electric charge. With the discovery of the 'neutron', the picture of the atom was complete: negatively charged electrons orbit an ultra-compact nucleus which contains protons and neutrons and accounts for 99.9 per cent of an atom's mass (the exception is an atom of the lightest element, hydrogen, which contains a lone proton in its nucleus).

Chadwick's discovery had crucial implications for a star more massive than the 'Chandrasekhar limit' of 1.4 solar masses. Yes, its innards would be turned to marshmallow and it would be crushed ever smaller by the merciless grip of gravity. But this was not the whole story. The runaway shrinkage of the star would, inevitably, squeeze the electrons into the nuclei, where they would react with the protons to make neutrons.

Neutrons, like electrons, are fermions. And a neutron gas, just like an electron gas, would make the star stiff enough to resist gravity. But neutrons are much smaller than atoms. Instead of a white dwarf the size of the Earth, the result would be a ball of neutrons the size of Mount Everest. So dense would be such a 'neutron star' that a sugar-cube-sized volume would weigh as much as the entire human race.

In the 1940s, the British astronomer Fred Hoyle would suggest that the only possible power source of a 'supernova' – a type of stellar explosion so bright it can often outshine a galaxy of 100 billion stars – is the gravitational energy released in the catastrophic shrinkage of a star to form a neutron star. But it was not until 1967 that a Cambridge graduate student, Jocelyn Bell, discovered a neutron star, in the guise of a rapidly spinning 'pulsar'.[13]

Although 'neutron degeneracy pressure' makes neutron stars stable against further gravitational collapse, such suns have the same Achilles' heel as white dwarfs. They are 'relativistic stars' whose constituent particles are flying about at close to the speed of light. Consequently, above a certain threshold of mass, even the stuff of a neutron star turns to marshmallow.

The physics of neutrons, which are held together by nature's 'strong nuclear force', is more complicated than the physics of electrons, which interact via the electromagnetic force. For this reason the threshold mass for a neutron star is not as precisely known as the Chandrasekhar limit. It was first calculated by the Russian physicist Lev Landau in 1932 and it is widely believed to be about three times the mass of the Sun. For stars above this mass, there is no known force that can stop their shrinkage down to a singularity.

This mass limit would not matter if there are no stars heavier than three times the mass of the Sun. But there most certainly are. A few rare ones are more than 100 times as massive as the Sun. Such stars are inherently unstable and prone to violent convulsions that eject large quantities of their mass during their lifetime. But, even taking this into consideration, they are still likely to be more massive than three solar masses when their internal fires finally flicker and go out. Runaway collapse to form a black hole appears unavoidable.

In fact, we know it is unavoidable. In 1971, NASA's 'Uhuru' satellite discovered the first black hole candidate: Cygnus X-1. A couple of dozen stellar-mass black holes are now known in our Galaxy. In addition, NASA's Hubble Space Telescope has confirmed that pretty much every galaxy in the Universe has a giant black hole lurking in its heart. Some are as big as 50 billion times the mass of the Sun whereas the one 27,000 light years away in the heart of the Milky Way – Sagittarius A* – is a mere 4.3 million times the mass of the Sun. The origin of such 'supermassive' black holes is one of the outstanding puzzles of modern astrophysics.

But black holes – which conjure singularities into the very heart of general relativity – are not the only problem for Einstein's theory of gravity. There is another one: the big bang.

The big bang

The general theory of relativity is a recipe for how matter – or, more generally, energy – warps the fabric of space-time. Einstein was never one to shy away from the really big problems in science. So in 1917 he applied his theory to the biggest collection of matter he could think of: the entire Universe.

Gravity orchestrates the large-scale Universe because mass comes in only one type, which always attracts. So, despite being by far the weakest of nature's fundamental forces, its effect builds remorselessly with increasing mass and, even on the scale of the planets, becomes irresistible, overwhelming all of nature's other fundamental forces. 'Gravity is a habit that is hard to shake off,' as Terry Pratchett put it.[14] By contrast, nature's 'strong' and 'weak' nuclear forces have ultra-short ranges, and the electromagnetic force, despite having an infinite range like gravity, is cancelled out on the large scale because of the existence of two types of electric charge which permit it to both attract and repel.

With its desire to pull everything together, gravity, like some cosmic Cupid, continually strives to break the terrible isolation of matter. From the beginning of time, when matter was blasted to the four corners of the Universe by the explosion of big bang, it has truly been nature's lonely hearts club force. 'Love', as Dan Simmons observed, 'was hardwired in the structure of the Universe as matter and gravity.'[15]

In applying his theory of gravity to the entire Universe, Einstein created 'cosmology', the science of the origin, evolution and ultimate fate of the Universe. But he went a little wrong. Like Newton before him, he believed the Universe was the way it had always been and the way it always would be. The great appeal of such an unchanging, or 'static', Universe was that it had no beginning or end so there was no need to waste time worrying about the sticky question of how it all got started.

The trouble was that Einstein's equations appeared to describe a dynamic space-time that was desperate to be in motion. Einstein fixed this by postulating that empty space contains energy, giving it intrinsic curvature, independent of any matter. This curvature, which he called the 'cosmological constant', manifests itself as a repulsion of empty space. So, although all the objects in the Universe pull on each other with the attractive force of gravity, this is perfectly counterbalanced by the repulsive force of empty space. Hey presto: a static Universe.

It was Einstein's greatest acolyte, Arthur Eddington, who showed in 1930 that Einstein's static Universe could never have worked. Like a pencil standing vertically on its point, it was unstable, and the slightest disturbance would cause it to veer from its balance point. The Universe envisaged by Einstein teetered on the knife edge between expansion and contraction. The merest of nudges would send it careering one way or the other.

But though Einstein missed the message in his own equations – that the Universe must be in motion – others did not. In order to simplify his equations so that they could be 'solved', Einstein had insisted that the density of matter in the Universe remains constant for all time. But the same year he made this assumption, 1917, Willem de Sitter – the Dutch recipient of one of the first smuggled-out copies of Einstein's theory – also applied the general theory of relativity to the Universe. And in marked contrast to Einstein, he did not insist that the density of matter remain constant but instead kept an open mind. And what de Sitter discovered was a Universe permitted by Einstein's theory whose space was expanding. If two test particles were put in such a Universe, the general expansion of space would steadily increase the distance between them.

The trouble with de Sitter's Universe was that it was empty. It was nothing but expanding space-time. It did not describe the Universe we live in. (Also, shockingly, it revealed the genie Einstein had let out of the bottle: space-time is a dynamic thing that can exist entirely independently of matter.)

But, in 1922, a Russian astronomer called Aleksandr Friedmann discovered a whole class of universes permitted by Einstein's theory that were either expanding or contracting and contained matter. Friedmann's 'evolving' universes were independently discovered five years later by a Belgian Catholic priest called Georges Abbé Lemaître. Most people today know of Friedmann–Lemaître universes by their more common moniker: 'big bang universes'.[16]

Friedmann and Lemaître's universes were, of course, merely theoretical. But everything changed in the 1920s because of an American astronomer called Edwin Hubble. For his first trick, he discovered 'galaxies'.

Einstein and others had been handicapped by not knowing

the true building blocks of the Universe. At the beginning of the twentieth century, it was known that the Sun belonged to a giant collection of stars called the Milky Way. Scattered about the sky were also myriad other fuzzy 'spiral nebulae'. The question was: were they clouds of luminous gas within the Milky Way or other islands of suns, or 'galaxies', so far beyond the Milky Way that sheer distance had blurred their stars into a fog?

In 1923, using the world's biggest 'eye', Hubble answered the question. He pointed the 100-inch Hooker Telescope on Mount Wilson in Southern California at the Great Nebula in Androm-eda. And not only was he able to see individual stars but stars of a special type whose regular brightening and fading revealed their distance. These 'Cepheid variables' showed beyond any doubt that Andromeda – and by inference all the spiral nebulae – were far, far beyond the Milky Way.[17]

Hubble had discovered the fundamental building blocks of the Universe: galaxies. Our Milky Way, with its 100 billion stars, was but one galaxy among about 2 trillion others.[18]

For his next trick, Hubble began measuring the speed at which galaxies were moving, continuing work begun on Mount Wilson by a former mule driver called Milton Humason.[19] By 1929, Hubble had measured the speed of enough galaxies to make an extraordinary discovery. Pretty much all of the galaxies were flying away from the Milky Way, hardly any were approaching. And the further away they were the faster they were receding. Hubble had discovered that the Universe is expanding. Remark-ably, the big bang solutions of Einstein's theory of gravity dis-covered by Friedmann and Lemaître describe reality.

But it was one thing to discover the expanding Universe and quite another to take on board what it meant. That involved taking the discovery seriously, and scientists always have im-mense trouble truly believing that their esoteric mathematical equations actually describe the real world.

But, in the late 1930s, and for entirely the wrong reason, the Ukrainian-American physicist George Gamow began thinking about the implications of an expanding Universe. The reason was that he was looking for a furnace to build up nature's elements.

There are ninety-two naturally occurring elements, ranging from the lightest, hydrogen, all the way up to the heaviest, uranium. Gamow believed the Universe had started out with hydrogen – think of it as nature's fundamental Lego brick – and all the other elements had been built up, step by step, from hydrogen. But this required a furnace at a temperature of many billions of degrees.[20]

Gamow believed stars did not fit the bill (he was wrong).[21] So he looked for another furnace. And that was when he began imagining the expansion of the Universe running backwards like a movie in reverse. After billions of years – we now know the time to be 13.82 billion years – all the matter of the Universe would be squeezed into the tiniest of tiny volumes. This was the moment of the Universe's birth: the big bang.

When something is squeezed into a small volume it gets hot, as anyone who has squeezed air into a bicycle pump knows. The big bang was therefore a hot big bang, Gamow realised. It would have been like the fireball of a nuclear explosion.

Gamow's furnace was incapable of building all of nature's elements.[22] But, in being wrong – and this is sometimes the case in science – Gamow was right. That the Universe was expanding implied that it had been born in a searing hot fireball. And when Gamow thought about the big bang fireball, he realised something extraordinary: its heat should still be around today.

The heat and light of a normal explosion – say, the explosion of a stick of dynamite, or even of a nuclear bomb – dissipates into the environment. After an hour, a day, a week, it is completely gone. But the Universe, by definition, is all there is. The heat of the big bang fireball had nowhere to go. It was bottled up in the Universe. Consequently, it must still be around today, greatly cooled by the expansion of the Universe in the past 13.82 billion years. Instead of appearing as high-energy visible light, it should appear as low-energy radio waves. In fact, Gamow's calculations revealed that this 'afterglow' of the big bang would account for 99.9 per cent of the particles of light, or 'photons', in the Universe.

But every physicist, be they Einstein or anyone else, gets so far and then makes a mistake. And Gamow's mistake was to think that the afterglow of the big bang had no features that would make it easy to recognise in today's Universe. His two students, however, realised that it did. Ralph Alpher and Robert Herman guessed it would have two very striking characteristics: one, it would be coming equally from every direction in the sky, and, two, and somewhat more technically, it would have the 'spectrum of a black body'.[23]

Alpher and Herman published their prediction in the inter national science journal *Nature* in 1948. But no one took any notice. Even worse, when they asked radio astronomers whether the afterglow of the big bang was detectable, they were told (incorrectly) that it was not.

Fast forward to 1965. Two radio astronomers at the American phone company AT&T had inherited a giant 'radio horn' at Holmdel, New Jersey. The horn had been used for pioneering experiments with the first communication satellites, 'Echo 1' and 'Telstar'. Arno Penzias and Robert Wilson wanted to do some astronomy with it. But everywhere in the sky they pointed the horn, they were getting a persistent hiss of static.[24]

First they thought the source was New York City, just over the horizon, but the hiss remained even when they pointed the horn in the opposite direction. They thought it might be a source in the Solar System but, as the months went by, and the Earth went round the Sun, the hiss did not change as expected. They thought the culprit could be an atmospheric nuclear test that had recently injected electrons high into the atmosphere, which would generate radio waves, but, as time passed, the hiss did not decline as expected.

Eventually, the gaze of Penzias and Wilson alighted on two pigeons which had nested inside the giant radio horn. They had coated the interior with a 'white dielectric material', more commonly known as pigeon droppings. Could this be the source of the spurious hiss of radio waves? Penzias and Wilson captured the pigeons and cleaned out the interior of their horn. But, frustratingly, the anomalous hiss remained.

Finally, Penzias learnt from a colleague of a search being made from nearby Princeton University for the heat relic of the early Universe. Incredibly, he and Wilson, totally by accident, had stumbled on the most important cosmological discovery since Hubble's discovery of the expanding Universe: the leftover heat from the birth of the Universe. They had confirmed the big bang.

It was one of the greatest discoveries in the history of science. The Universe had not existed for ever. It had been born. There was a day without a yesterday. For the discovery of the 'cosmic background radiation', Penzias and Wilson would win the 1978 Nobel Prize for Physics.

The arrow of time

One of the mysteries of our Universe is why time flows in the direction it does. Why do people grow old, eggs break and castles crumble but we never see people grow young, eggs unbreak and castles uncrumble? For the answer it is necessary to go back to the big bang.

All the above changes are associated with transformations from order to disorder. But there are many ways an egg can be broken (disordered) and only one way it can be intact (ordered). And since all possibilities are equally likely, it is overwhelmingly likely that an egg will go from being intact to being broken. This is the 'Second Law of Thermodynamics', which says that 'entropy' (microscopic disorder) can only increase. It is not impossible that an egg will go from broken to intact but it is overwhelmingly improbable.

But if the direction of time is associated with the Universe becoming more disordered, it implies that the Universe must have been very ordered in the past – that is, in the big bang. This poses a problem for physicists since an ordered state is an unlikely state. This is where gravity may come to the rescue, according to Larry Schulman of Clarkson University in New York.[25]

Initially, the Universe was a fireball in which matter was spread uniformly. This was a disordered state. But around 380,000 years after the moment of creation, the fireball had cooled enough for electrons to combine with nuclei to make the first atoms.

Free electrons interact strongly with photons whereas electrons in atoms do not – and there were about 10 billion photons for every electron. Consequently, before the formation of atoms, matter was blasted apart by photons and gravity could not pull it together into clumps. Afterwards, it could. It was as if gravity had suddenly 'switched on' in the Universe. Those clumps grew and grew and, ultimately, became the clusters of galaxies we see around us today.

For material experiencing gravity, the most likely state is with it clumped into objects like galaxies and stars. But, as pointed out, when the Universe was 380,000 years old, the matter of the Universe was spread uniformly in a very unlikely state. But the 'switching on' of gravity instantly transformed the Universe into an unlikely, special state, exactly as required if the arrow of time is to go the way it goes.

What is remarkable about this explanation is that the Universe looked pretty much the same just before and just after 380,000 years – the 'epoch of last scattering'. All that happened was that gravity went from being impotent to all-powerful. But from the point of view of gravity, the Universe went from being in a likely state to an unlikely state. A similar argument to Schulman's has been proposed by the British physicist Roger Penrose.

Penzias and Wilson's discovery of the afterglow of the fireball of the big bang created problems – lots of problems. If the Universe had begun in a big bang, what was the big bang? What had driven the big bang? And what had happened before the big bang? Nobody had wanted to face such questions, which was why most astronomers, including Penzias and Wilson, had subscribed to an eternal Universe known as the 'Steady State' theory.

But there was one difficulty that went to the very heart of general relativity. If the expansion of the Universe were imagined running backwards, as Gamow had pictured it, it would get ever denser, ever hotter and the space-time would become ever more curved. In fact, everything would skyrocket to infinity. It was another monstrous singularity. A singularity in time rather than in space, as in a black hole, but a singularity nonetheless.

So now there were two places where Einstein's theory broke

down, not just one. General relativity, far from being a perfect garment, was revealing itself to be a moth-eaten cloak.

But all hope was not lost for Einstein's theory.[26] Its singularities were not inevitable. One way out remained.

Singularity theorems

When relativity turns a dying star to marshmallow, the marshmallow is unlikely to be perfectly smooth. There is bound to be a lump here, a lump there. And as the star is crushed ever smaller by gravity, this unevenness will be magnified. In other words, the star will not shrink perfectly symmetrically. What this means is that different parts of the collapsing star may not, in the end, all pile up at one impossibly dense point. They may miss each other. There will be no singularity. And Einstein's theory of gravity will live to fight another day.

And what is true of black holes might also be true of the big bang. If the matter of the Universe is spread unevenly, this unevenness would become magnified as the backward-running Universe shrank ever smaller. Different parts of the collapsing Universe, instead of all piling up at one point, would miss each other and so not create a catastrophic singularity. Since Einstein's theory of gravity would not break down, it would be possible to use the recipe to follow the history of the Universe to an earlier time before the big bang. Perhaps, for instance, the Universe had contracted down to a big crunch from which it had then bounced in the big bang.

Enter English theorists Stephen Hawking and Roger Penrose. Between 1965 and 1970, the question of whether the singularities in the big bang and in black holes could be avoided became the focus of their research work. And the pair proved a range of powerful 'singularity theorems'. The most important of them showed that under a wide range of general and highly plausible conditions the singularities in the big bang and black holes were unavoidable. They formed no matter how the backward-running movie of the Universe went, no matter how a star shrank to make a black hole.

There was no escaping the inconvenient truth. All along, Einstein's theory of gravity had contained the seeds of its own destruction. Although it correctly predicts light bending, the precession of the perihelion of Mercury and the slowing of time in strong gravity, it also predicts singularities, which are nonsensical. It breaks down in the heart of black holes and at the beginning of time. 'If we can't understand what happened at the singularity we came out of, then we don't seem to have any understanding of the laws of particle physics,' says Neil Turok of the Perimeter Institute in Waterloo, Canada.

The breakdown of Einstein's theory of gravity at singularities can only mean that it is an approximation of a better, deeper theory.

Quantum gravity

The two towering achievements of twentieth-century physics are Einstein's theory of gravity – the general theory of relativity – and quantum theory.[27] Each has passed every experimental and observational test with flying colours and each, in its own domain, reigns supreme. General relativity is a theory of big things like stars and the Universe while quantum theory is a theory of small things like atoms and their constituents.[28] But near a singularity – either in a black hole or in the big bang – something big is squeezed into a volume smaller than an atom. So, in order to understand what happens at the heart of black holes and, most importantly, shed light on the origin of the Universe, it is necessary to unite general relativity with quantum theory to create a 'quantum theory of gravity'. This is the name given to the deeper theory, desperately being sought by physicists.

As early as 1916, Einstein recognised that quantum theory, if it was nature's last word – and he did not believe that it was – would require a modification of general relativity. 'Due to the inner-atomic movement of electrons, atoms would have to radiate not only electromagnetic but also gravitational energy, if only in tiny amounts,' he wrote. 'As this is hardly true in Nature, it appears that quantum theory would have to modify not only Maxwellian electrodynamics, but also the new theory of gravitation.'[29]

Some idea of the extreme difficulty in finding a quantum theory of gravity can be appreciated by understanding the bizarreness of quantum theory and how fundamentally different it is from general relativity . . .

Further reading

Fölsing, Albrecht, *Albert Einstein*, Penguin, London, 1998.

Levenson, Thomas, *Einstein in Berlin*, Bantam Books, New York, 2003.

Levenson, Thomas, *The Hunt for Vulcan . . . And how Albert Einstein destroyed a planet, discovered relativity and deciphered the Universe*, Head of Zeus, London, 2015.

Miller, Arthur, *Empire of the Stars: Friendship, betrayal and obsession in the quest for black holes*, Little, Brown, London, 2005.

PART THREE

Beyond Einstein

A quantum of space-time

*How quantum theory implies that space and
time are doomed and must somehow emerge
from something more fundamental*

On Mondays, Wednesdays and Fridays, we teach the wave
theory and on Tuesday, Thursdays and Saturdays the particle theory.

William Bragg[1]

Your theory is crazy but is it crazy enough to be true?

Niels Bohr[2]

Quantum theory is fantastically successful. It has given us
lasers and computers and nuclear reactors, and explains why
the ground beneath our feet is solid and the Sun shines. But, in
addition to being a recipe for understanding things and building
things, quantum theory provides a unique window on a counter-
intuitive, Alice in Wonderland world just beneath the skin of
reality. It is a place where a single atom can be in two locations
at once – the equivalent of you being in New York and London
at the same time; a place where things happen for absolutely no
reason at all; and a place where two atoms can influence each
other instantaneously even if on opposite sides of the Universe.

The need for quantum theory arose out of Maxwell's theory
of electromagnetism, which describes all electrical and magnetic
phenomena in one elegant and seamless framework. The theory
actually contains not one but two paradoxes, both of which
involve light. The resolution of the first – how there can be a

unique speed of light in a vacuum, independent of the speed of any observer – led to one of the great revolutions in twentieth-century physics: Einstein's special theory of relativity. The resolution of the second paradox led to the other great revolution: quantum theory.

The second paradox arises because Maxwell's theory permits electromagnetic waves of any size. So, in addition to visible light, which has a 'wavelength' of just under a thousandth of a millimetre, it is possible to have waves with a longer wavelength such as radio waves – discovered by Heinrich Hertz in 1888 – and waves of shorter wavelength such as X-rays – discovered by Wilhelm Röntgen in 1895. The size of the wave is related to the energy it carries: sluggish radio waves are far less energetic than waves of visible light which in turn are far less energetic than rapidly oscillating X-rays.

In a hot gas of atoms, light waves are repeatedly emitted and absorbed, and the result, if enough time passes, is the creation of all possible light waves. In such a state of 'thermal equilibrium', the energy is shared out equally among all the wavelengths. But herein lies a problem. Although there is a limit on how long the wavelength of light can be – set by the size of any container – there is no corresponding limit on how short the wavelength can be. This means that if we pick a wavelength – any wavelength whatsoever – there will always be a finite number of waves with longer wavelength but, crucially, an infinite number with shorter wavelength.

As pointed out before, the energy must be shared out equally among all the waves. Since there are hugely more waves with shorter wavelengths than longer wavelengths, this means that most of the energy will always be carried by the shorter waves. Inevitably, then, all the energy in a hot gas will end up in the highest-energy X-rays. Before the discovery of X-rays in 1895, the highest-energy light known was ultraviolet, which was why this result became known as the 'ultraviolet catastrophe'.[3]

The paradox is put in its most dramatic form by considering our Sun. Theory predicts that our local star should instantly radiate away all of its heat in a blinding flash of X-rays. So how is it

that it is still shining? 'There is hardly any paradox without util-
ity,' wrote the German mathematician Gottfried Leibniz. And by
finding the revolutionary answer in 1900, the German physicist
Max Planck proved him right.

Quanta

The 'killer app' for electricity in the late nineteenth century was
the light bulb. A question of key technical and economic importance was therefore: How do you maximise the visible light given
out by the heated filament of a light bulb? There was obviously
no chance of answering such a question while the best theory of
light predicted that a hot filament – just like the hot gas of the
Sun – should instantly radiate all its light in a flash of X-rays.

What was needed was a way to tame light and so avoid the
nonsensical scenario of the ultraviolet catastrophe. And Planck,
after a great deal of head-scratching and mental torment, finally
found one.

According to Maxwell's theory, an oscillating electric charge
such as an electron radiates light at its oscillation 'frequency'.
Actually, the theory says that an accelerated charge broadcasts
electromagnetic radiation, but an oscillating charge is simply
one that is repeatedly accelerating. So, Planck imagined a container whose walls are made of electrons attached like weights to
springs. Nowadays, of course, we know that Planck's oscillating
electrons exist inside atoms but, at the end of the nineteenth
century, many physicists still doubted the existence of atoms.
Planck's picture of electrons on springs is good enough, however.

If the container is heated, the heat energy makes the springs os-
cillate, and the oscillating springs produce oscillating light waves
at exactly the same frequency. These waves cross the container and
are absorbed by other oscillating springs, which in turn produce
oscillating light waves at their own frequencies. And the result of
all the countless interactions is that the heat-energy is shared out
equally between all the springs and light waves. This is the situation in which the highest-frequency light waves get most of the
energy because they are overwhelmingly more common.

Planck realised that this catastrophe can be tamed if the oscillating springs are not free to give out or absorb any amount of energy whatsoever but are instead restricted to giving out or absorbing energy at only multiples of a basic amount. The amount, he proposed, was h times their frequency, f, where h was a very small number (frequency is defined as the number of oscillations per second).

Think how ridiculous this is. It is like a high-jumper being able only to jump heights that are multiples of, say 0.5 metres. In other words, they can jump 0.5 metres, or 1.0 metre, or 1.5 metres. But they cannot in any circumstances jump a height of 0.75 metres or 1.2 metres or 1.81 metres.

There was no plausible reason why Planck's atom-springs should only give out energy in multiples of hf. His scheme was utterly mad. He came up with it for one reason and one reason only: it worked. It correctly predicted the way in which the amount, or intensity, of light from a gas of hot atoms varied with frequency, or equivalently energy.

According to Planck, an oscillator cannot simply absorb light and then emit light at a slightly higher energy. It can emit light at only the next highest energy permissible. It is an all-or-nothing thing. If the oscillator does not have enough energy to make the light, the light simply does not get made. So, when the energy gets shared out between the light waves, crucially, the highest-frequency waves do not get the lion's share of that energy, or even any energy at all. They are simply too energy-expensive. With the highest-energy light tamed in this way, there is no ultraviolet catastrophe.

The paradox of travelling alongside a light beam and seeing something impossible arose because Newton's theory imposed no limit on the speed of a body. The paradox of the ultraviolet catastrophe arose because Maxwell's theory imposed no limit on the smallness of the wavelength of light. Just as Einstein's finite speed-of-light limit tamed the infinite, Planck's quantum tamed the infinitesimal.

To Planck, his scheme was nothing more than a mathematical fudge. Though he claimed that energy was absorbed by atoms in discrete chunks, or 'quanta', with energy always a multiple of hf, he did not for a moment think that light actually flew through space in this way. That claim was left to Einstein, who spawned two revolutions: relativity and quantum theory. In his 'miraculous year' of 1905, he wondered about the striking similarity between Planck's formula for the spread of energy among the different wavelengths of light in a container and Maxwell's formula for the spread of energy among the particles of a gas.

Maxwell was a genius who, despite dying at the tragically early age of forty-eight, made key contributions not only to electromagnetism but also to astronomy and the microscopic theory of gases. To obtain his formula for the distribution of energy among particles of a gas, he imagined atoms flying about like tiny bullets and worked out how countless collisions between them, each of which transferred energy from fast-moving to slow-moving particles, shared out the total energy. The striking similarity between Maxwell's formula and Planck's formula, Einstein reasoned, could mean only one thing: light too consists of bullet-like particles. What Planck had considered to be no more than a mathematical sleight of hand was reality. Light really is emitted and absorbed in particle-like chunks, later christened 'photons'.

We now know that everything comes in indivisible chunks, or quanta: energy, matter, electric charge, and so on. On the smallest scales nature is not continuous, as classical physics imagined, but grainy, like a newspaper photograph inspected close-up.

The 'physical constant' h became known as Planck's constant. Because it is extremely small, the energy carried by a single photon is minuscule and so we never notice that the light from a light bulb is in fact a torrent of tiny bullets. There are simply too many of them.

To visualise what h does to the microscopic world, imagine it is possible to make it bigger and bigger until its consequences become apparent in the everyday world. Eventually, individual photons carry so much energy that the filament of the light bulb can create only small numbers of them. So it starts flickering. One moment it makes 10 photons, the next 7, the next 15, and so on. If h is made bigger still, photons become so energy-expensive that the filament cannot make even a single photon, and the bulb stops stuttering and goes dark.

Einstein used the idea that light consists of photons to explain a puzzling phenomenon – the ejection of electrons from the surface of certain metals.[4] His explanation of the 'photoelectric effect' not only earned him the 1921 Nobel Prize for Physics but was the only work Einstein himself considered to be 'revolutionary'.[5] The reason can be appreciated from a remarkably ordinary, everyday observation . . .

Random reality

Look out of a window. You will see the scene outside and, if you look closely enough, a faint reflection of your face as well. This is because glass is not perfectly transmitting. Although most of the light that strikes it goes through, a small amount bounces back.

What happens at a window is easy to explain if light is a wave. Think of a wave spreading across a lake and encountering an obstacle – say, a submerged log. Most of the wave continues on while part of it is turned back. But what happens at a window is not easy to explain if light is a stream of photons, all of which are the same. After all, if they are all identical, surely they should be affected identically by the window (we are assuming perfect, flawless glass, by the way!). Either all should go through or all should be reflected. There is no way that most can go through while some bounce back.

To explain why you can see your face in a window, physicists had no choice but to water down their definition of 'identical'. For photons, identical must mean only that they have the same 'chance' of being transmitted by the glass – say 95 per cent – and the same 'chance' of being reflected – say 5 per cent. But the introduction of the word chance into physics, as Einstein realised, is catastrophic.

Physics is a recipe for predicting the future with 100 per cent certainty. If the Moon is at a particular location today, its location tomorrow can be predicted with 100 per cent certainty by using Newton's law of gravity. But seeing your face in a window tells us that it is impossible to predict with certainty what an individual photon will do on encountering a window pane. It is possible to predict only its 'probability' of being transmitted or reflected.

Think for a moment what this means. If you roll a dice, you may think the outcome is unpredictable. But, actually, if you knew the exact velocity with which the dice was rolled, the motion of the air currents in its vicinity, and so on, it would be possible, with the aid of a big enough computer, to predict the number that comes up. Everything we think of in the everyday world as random is not really random. It is unpredictable only in practice. In marked contrast, what a photon does when it strikes a window pane is unpredictable in principle. It would not matter how much information was available, or how big a computer was used, the photon's course of action would never be 100 per cent predictable. For the quantum dice, every throw is always the first throw.

And what is true for photons is also true for all the other microscopic building blocks of the world – from electrons to quarks. Every last one of them behaves in a manner which is fundamentally unpredictable.

How, then, is the everyday world predictable? The Sun will come up tomorrow morning and a ball thrown through the air will follow a trajectory predictable enough that it is possible to catch it. The answer is that what nature takes with one hand it grudgingly gives back with the other. Although the world is fundamentally unpredictable, its *unpredictability is predictable*. And the recipe for predicting the unpredictability is 'quantum theory'.

The revelation that the Universe is ultimately founded on random chance is arguably the single most shocking discovery in the history of science. And the remarkable thing is that it stares you in the face every time you look through a window. Einstein so hated the idea that he famously said: 'God does not play dice with the Universe.' Niels Bohr, the quantum pioneer, retorted: 'Stop telling God what to do with his dice.'

Einstein was not only wrong but spectacularly wrong. Not only does God play dice but, if He did not, there would not be a Universe – or at least a Universe of the complexity necessary for us to be here.[6]

Wave–particle duality

Seeing your face reflected in a window can be understood if light is a wave and it can also be understood if light is a stream of particles. In fact, this wave–particle duality is a key feature of the microscopic world of atoms and their constituents.[7]

Particles, which are localised, and waves, which are spread-out, would appear to be fundamentally incompatible. Certainly, that was the view of the physicists of the 1920s, who picked up the ideas of Planck and Einstein and ran with them. 'I remember discussions which went through many hours until very late at night and ended almost in despair,' wrote German physicist Werner Heisenberg. 'When, at the end of the discussion, I went alone for a walk in the neighbouring park I repeated to myself again and again the question: Can nature possibly be so absurd as it seemed to us in these atomic experiments?'[8]

The answer is yes. The microscopic world of atoms and their constituents is utterly unlike the everyday world (since it is a billion times smaller, perhaps we should never have expected it to be the same). Photons and their microscopic compatriots are neither particles nor waves but something else for which we have no word in our vocabulary and nothing in the everyday world around us to compare them with. Like shadows of an object we cannot see, we are limited to seeing particle-like shadows and wave-like shadows but never the thing itself. 'It has been possible to invent a mathematical scheme [quantum theory] . . . which seems entirely adequate for the treatment of atomic processes,' said Heisenberg. 'For visualisation, however, we must content ourselves with two incomplete analogies – the wave picture and the corpuscular picture.'

OK. So the fundamental building blocks of the Universe behave as particles and waves. But the waves are decidedly odd.

They are mathematical 'waves of probability' that encapsulate the chance of finding a particle anywhere or of it doing something. The probability wave spreads throughout space, bouncing off obstacles and 'interfering' with itself.[9] And the way in which it spreads is described by the 'Schrödinger equation', formulated by the Austrian physicist Erwin Schrödinger in 1925. At locations where the wave is big – or, specifically, has a large 'amplitude', there is a high probability of finding a particle, and at places where it is small a small probability.[10]

The genius of the Schrödinger equation, which Schrödinger actually guessed while on a weekend skiing trip with an old girlfriend, is that it unites the wave-like and particle-like facets of nature. It is the mathematical machinery that makes the wave–particle duality of the world concrete and permits physicists to calculate things in the real world. The same year Schrödinger guessed his equation, Heisenberg, together with Max Born and Pascual Jordan, invented a superficially different, but entirely equivalent, version of quantum theory called 'matrix mechanics'.

Multiple realities

Wave–particle duality is a two-way street. In 1923, the French physicist Louis de Broglie made the claim that not only can light waves behave like localised particles but particles such as electrons can behave as spread-out waves. It seemed mad. But, in 1927, Clinton Davisson and Lester Germer in the US and George Thomson in Scotland demonstrated that electrons could interfere with each other, their quantum waves reinforcing and cancelling each other out just like ripples overlapping on a pond. The irony is that George Thomson's father was 'J. J.' Thomson, who had discovered the electron. The father won the Nobel Prize for proving that the electron is a particle; the son won the Nobel Prize for showing that it isn't.

If the consequence for physics of waves behaving as particles is shocking, the consequence of particles behaving as waves is equally shocking. The reason is that the fundamental building blocks of matter can do all the myriad things that waves can do. Although those things have mundane consequences in the everyday world, they have earth-shattering consequences in the microscopic world.

Picture big waves at sea whipped up by a storm. Now picture the scene when the storm has passed and the water is merely ruffled by a gentle breeze. Anyone who has seen both types of wave will know that it is possible also to have a combination of the two – a big rolling wave whose surface is gently rippled. And this, it turns out, is typical not only of water waves but of all waves. If two waves exist, then a combination, or 'superposition', of those waves can also exist. It seems a trivial observation. But in the submicroscopic world it is far from trivial.

Say there is a quantum wave that represents an oxygen atom (the technical name for such a probability wave is the 'wave function'). Say it is highly peaked on the left-hand side of a room. In other words, there is pretty much a 100 per cent chance of finding the atom on the left-hand side. Now imagine a quantum wave for the oxygen atom that is highly peaked on the right-hand side of a room. So there is almost a 100 per cent chance of finding the atom on the right hand side. Nothing remarkable here. But, remember, if two waves are possible, so too is a superposition of the two. However, a superposition of the two quantum waves corresponds to an oxygen atom that is simultaneously on the left-hand side of the room and the right-hand side of the room – in two places at once.

But nobody ever observes an oxygen atom in two places at once.[11] If the oxygen atom is found on the left-hand side of the room, then the wave representing the oxygen atom on the right-hand side of the room instantly 'collapses'. This is what the Schrödinger equation tells us. Until an observation is made there exists a haze of possibilities, but the moment an observation is made one possibility and one possibility only is actualised so that a particle exists in a particular location with 100 per cent certainty. The triumph of the Schrödinger equation is that it reconciles the apparently irreconcilable, encapsulating both the wave-like and particle-like faces of nature in one mathematical expression.[12]

But, if nobody ever observes an oxygen atom – or anything else, for that matter – in two places at once, who cares about the phenomenon of quantum-wave superposition? The answer is that it has *consequences*. And those consequences lead to all kinds of quantum weirdness.

Here is a simple example. Two identical bowling balls collide and ricochet. They fly outwards from the collision point in opposite directions. Now say they collide over and over and you note the direction they travel outwards. Say, towards 2 o'clock and 8 o'clock, 4 o'clock and 10 o'clock, and so on. After repeating this hundreds of times it will be obvious that the bowling balls have flown off to every point on a clock face, in every possible direction.

Picture doing the same thing with two identical quantum objects such as two electrons or two oxygen atoms. After colliding them hundreds of times it will be obvious that there are some directions where the quantum particles never go – say, 3 o'clock and 9 o'clock, and 5 o'clock and 11 o'clock. Why? Because these are the directions in which the peaks of the probability wave for one particle coincide with the troughs of the probability wave for the other. So they cancel each other out, or destructively interfere, leaving a probability of zero of finding the particles.

The point is that 'interference' enables two quantum waves in a superposition to interact before a quantum particle is observed. And this can have unexpected consequences – like colliding particles never scattering off each other in particular directions.

It also explains why an electron orbiting in an atom does not fall into a nucleus, as Maxwell's theory indicates it should. There are a myriad possible paths that an electron could take as it heads for the nucleus. It could spiral in, or it could head in a straight line, or it could take a wiggly path, and so on. And associated with each is a quantum wave. But it turns out that, close to the nucleus, all the quantum waves destructively interfere, cancelling each other out, so that there is no probability of finding the electron there.

This highlights another fundamental difference between quantum physics and pre-quantum physics. In 'classical' physics, a body such as the Moon travels along a unique and well-defined trajectory. In quantum theory, there is no such thing as a well-defined trajectory. Between observations, an electron can be thought of as travelling along multiple paths, each of which has a particular probability associated with it.

But, if quantum properties such as superposition are not weird enough, they can combine to create even weirder quantum phenomena. 'Non-locality', or 'spooky action at a distance', for instance, was considered so mad by Einstein that he believed it proved quantum theory is not nature's final word but merely an approximation of a deeper theory. To appreciate it, it is necessary to know about 'spin'.

Faster-than-light influence

Quantum spin is another one of those quantum properties like wave–particle duality and unpredictability which has no analogue in the everyday world. Think of an ice skater spinning on the ice. She possesses a thing called 'angular momentum', which is simply her ordinary momentum multiplied by the average distance of her body from the axis she is spinning around. Angular momentum, like ordinary momentum and energy, is one of those quantities that can never be created or destroyed but is 'conserved'. This is why, if the ice skater pulls in her arms, bringing her body closer on average to the spin axis, she spins faster to compensate.

The quantum twist is that particles such as electrons behave as if they are spinning even though they aren't. They have intrinsic spin. And just like everything else in the submicroscopic world, it comes in indivisible quanta. For historical (and confusing) reasons, the fundamental unit of spin is ½ of a certain quantity (the quantity is $h/2\pi$). This is the spin carried by an electron. And it turns out there are only two possible ways it can spin. They can be thought of as clockwise and anticlockwise, although of course an electron is not actually spinning! Physicists prefer to refer to the two possibilities as spin 'up' and 'down'.

Here is how spin plus a few other quantum properties

– superpositions and unpredictability – lead to spooky action at a distance.

Take two electrons. The first electron can be spin up and the second spin down. Or the first can be spin down and the second spin up. But, crucially, it is possible to have a superposition in which the two electrons are spin up, down *and* spin down, up.

Now because the electrons have opposite spin, their spins cancel out – that is, their angular momentum is zero. But remember angular momentum can never change. So it must always remain zero – in other words, the spins of the two electrons must always be opposite.

Without looking at either electron, put one in a box and take it to a distant part of the globe. Now, open the box. Because of quantum unpredictability, the electron has a 50 per cent chance of spinning up when it is observed and a 50 per cent chance of spinning down. But – and this is the key – if the electron is observed to be up, the stay-at-home electron must instantaneously become down, and vice versa. Notice that word instantaneously. This is in total violation of Einstein's cosmic speed limit of the speed of light. Which is why Einstein thought this spooky action at a distance had to prove that quantum theory was incorrect.

Unfortunately for Einstein, laboratory experiments have shown that subatomic particles born together – like these two electrons – can indeed influence each other faster than light. Even if they are on opposite sides of the Universe. In the jargon, they are 'entangled'. As Niels Bohr said: 'If anybody says he can think about quantum physics without getting giddy, that only shows he has not understood the first thing about them.'

Non-locality, also known as 'entanglement', *is* compatible with special relativity as long as special relativity forbids the transmission of 'information' at speeds faster than light. In the case of the two electrons, you can never know whether an electron is up or down until you look at it, and then the direction it takes is random. So encoding a message – for instance, making up a '1' and down a '0' – can never work. All that can ever be transmitted is random gobbledygook, useless information, never a true message.

But, apart from unpredictability, superpositions and entanglement, there is an even more basic property of waves that has implications for reality . . .

The uncertainty principle

Think of a wave that undulates up and down with a constant wavelength. Such a 'sine wave' marches on for ever, which means its precise location is 100 per cent uncertain. Now think of the momentum carried by the wave. Intuitively, it is related to its wavelength, with a very wiggly wave – that is, one of short wavelength – carrying a lot of momentum, and a sluggish wave – one of long wavelength – carrying little momentum. Because the sine wave is a wave of only a single wavelength, it has a precise momentum. To labour the point, its momentum is 100 per cent known.

Now, it is always possible to create a wave that is more localised than a sine wave. To create such a 'wave packet', simply add another sine wave, with a different wavelength. And another. And another . . . It is also possible to arrange things so that the sine waves cancel out everywhere *except in a localised region*.[13] And the more waves that are superposed, the more localised the wave can be made. But – and this is the point – there is a price to pay for pinning down where the wave is. Since the wave is now composed of a number of sine waves, each with its own characteristic wavelength – and, crucially, its own characteristic momentum – the wave's overall momentum is now uncertain.

So, the cost of knowing the location of a wave more precisely is knowing its momentum less precisely. And the opposite is also true. Recall that, in the case of the single sine wave, it was possible to know its momentum with 100 per cent certainty but only at the cost of its location being 100 per cent uncertain. There is a trade-off between our knowledge of the location of a wave and of its momentum. And it is a fundamental property of all types of wave. There is no way of getting around it. And since the microscopic building blocks of matter behave like waves, they too are subject to the same trade-off between knowledge of their location and knowledge of their momentum. We have met it before. It is called the Heisenberg Uncertainty Principle.[14]

To be more precise, the product of the uncertainty in a particle's location and the uncertainty in its momentum cannot be smaller than $h/2\pi$.[15] And a similar constraint holds for energy and time. Specifically, the product of the uncertainty in a particle's energy and the uncertainty in the time for which it exists cannot be smaller than $h/2\pi$.[16]

It is because h is so very small and your momentum is so large that you do not behave as a spread-out wave with an uncertain location. But, for tiny subatomic particles with low momentum, the uncertainty in their location is great. The building block of everyday matter with the smallest mass, and therefore momentum, is the electron. It consequently exhibits the most marked wave properties with the most uncertainty in its location. In fact, as pointed out in Chapter 7, this is another way of understanding why atoms exist and their electrons do not spiral into their central nuclei. An electron cannot be squeezed into a small volume near the nucleus because, having the biggest quantum wave, it needs *loads of elbow room*.

The Heisenberg Uncertainty Principle is actually the protector of the quantum world. If a quantum entity is located too precisely, it no longer has the spread-out waviness which is critical for it to exhibit interference and all the other wave phenomena behind quantum behaviour.

The disintegration of space and time

The Heisenberg Uncertainty Principle has profound consequences for the empty space. It means that smaller and smaller regions of the vacuum have larger and larger uncertainties in energy contained within them. The energy pops into existence and pops out again like money stolen from a wallet which is returned before the owner notices its absence. Such 'quantum fluctuations' manifest themselves as particle–antiparticle pairs such as electrons and positrons appearing out of nothing like rabbits out of a hat. But they have such a fleeting existence, disappearing in the merest split-second whence they came, that it is actually a stretch to call them real particles. Nevertheless, such 'virtual' particles have a real effect on atoms, buffeting their outer electrons and causing a tiny change in the energy of the light those electrons give out as they jump between orbits. For measuring this 'Lamb shift' in the light of the hydrogen atom, the American physicist Willis Lamb won the 1955 Nobel Prize for Physics.

Because of quantum fluctuations, the vacuum is actually seething with energy. And on the smallest scales, where the fluctuations are large enough, that energy is sufficient grossly to warp space-time.[17]

Think of the vacuum as like the ocean on a stormy day. From the perspective of a high-flying seagull, the ocean looks perfectly smooth. This is what space-time looks like on the largest scales. But from the perspective of a seagull flying much lower, big rolling waves can be seen on the ocean. Similarly, on smaller scales, space-time begins to convulse. Finally, from the perspective of a seagull on the deck of a trawler, waves can be seen smashing over the bow. All is froth and chaos. And this is what space-time is believed to look like on the smallest possible scales.

John Wheeler coined the name 'quantum foam' for this chaotic space-time. But it should be stressed we currently have no observational evidence that it exists. Although quantum foam should by rights affect the light from distant events in the Universe such as 'quasars' and 'gamma ray bursters' on its multi-billion-year journey to Earth, no one has yet detected the effect.[18]

Most physicists agree with Wheeler that, on the smallest scales, space-time does not exist. 'Space-time is doomed – that much is pretty universally agreed,' says Nima Arkani-Hamed of the Institute for Advanced Study in Princeton, New Jersey. 'It must be replaced by more fundamental building blocks. The question is what exactly?'

Arkani-Hamed is widely considered one of the world's most original and talented theoretical physicists. With his trademark black T-shirt, shorts, sandals and long flowing black hair, he cuts a striking figure at a blackboard, scrawling equations and waving his arms wildly for emphasis. Generous with his time, he will talk physics with absolutely anyone. In fact, he claims he has never turned down a graduate student who wanted to work with him.[19]

That Arkani-Hamed is at the epicentre of twenty-first-century physics is somewhat of a miracle. Aged ten, he nearly died of a fever in the mountains between Iran and Turkey as his family fled the Khomeini regime in 1982. As his mother's horse carried him through the night, she kept him conscious by pointing out the phosphorescent band of the Milky Way and promising him a telescope when they got to safety. He duly got his telescope in Toronto, Canada, and eventually made his way via Berkeley in California and Harvard to the Institute for Advanced Study, famous for being the place where Einstein and logician Kurt Gödel spent their twilight days.

Arkani-Hamed is using his seemingly boundless energy and enthusiasm to persuade the Chinese to build a particle accelerator to dwarf the LHC and probe nature on a scale ten times smaller and an energy ten times higher than the European machine. If it comes off, the 'Great Collider' could be operational by 2042. But Arkani-Hamed's theoretical focus is firmly on finding a deeper theory than Einstein's theory of gravity. And because that theory recognises that gravity is nothing more than the curvature of space-time, the quest to understand gravity has been transformed into a quest to understand *the origin of space and time*.

One impossibly tiny-length scale is thought by physicists to be of particular significance. At a length of 1.6×10^{-35} metres – 10 million billion billion times smaller than an atom – the force of gravity becomes comparable in strength to the other three fundamental forces of nature: the electromagnetic, strong nuclear and weak nuclear forces. The 'Planck length' was even recognised by Planck in 1900, though not for modern reasons. He simply thought it was so fundamental a scale that it would 'retain its significance for all times and all cultures, even extraterrestrial and extra-human ones'.[20]

The non-gravitational forces have all been successfully described by quantum theory, which suggests that a quantum description of gravity may be required to understand what is happening at or near the Planck scale. In the quantum picture, the fundamental forces are a consequence of force-carrying particles which are exchanged like a tennis ball batted back and forth between tennis players. In the case of electromagnetism, the force carrier is the photon; in the case of the weak nuclear force, three 'vector bosons'; and, in the case of the strong nuclear force, eight 'gluons'. Since such force-carrying particles are 'virtual' – popping out of the vacuum and popping back again – the more mass-energy they contain the briefer their existence and the shorter the distance they can travel during that existence. This means that the more massive the force-carrying particle the shorter the range of the force it mediates. For instance, the massive vector bosons give the weak nuclear force a range much smaller than the span of an atomic nucleus whereas the zero-mass photon endows the electromagnetic force with an infinite range.

It follows that, if a quantum description of gravity is possible, there ought to exist a force-carrier which carries the gravitational force. Theorists have christened this hypothetical particle the 'graviton'. There are many theoretical problems with the graviton and it is possible that no such particle exists. For instance, the strength of a force is synonymous with how frequently the force-carriers interact with particles that 'feel' the force. But gravity's incredible weakness compared with the other forces – the gravity between a proton and electron in a hydrogen atom is 10,000 billion billion billion billion times weaker than the electromagnetic force – means that gravitons hardly ever interact with matter. In fact, a detector of the mass of Jupiter would need to wait more than the age of the Universe before its bulk could stop a graviton.[21]

Notwithstanding problems with the graviton, uniting Einstein's theory of gravity with quantum theory is likely to be very hard because the two theories appear fundamentally incompatible. For one thing, general relativity is a theory about certainty – it predicts the future with 100 per cent certainty – whereas quantum theory is a theory of uncertainty – merely predicting the probability of possible alternative futures. 'Despite this, however, physicists have succeeded in finding a quantum description of nature's other fundamental forces,' says David Tong of the University of Cambridge.

But quantum theory even denies the existence of precise locations in space and trajectories of bodies moving through space, which are the very foundation stones of Einstein's theory of gravity. Quantum theory also views the Universe on the smallest scales as discrete and grainy whereas general relativity sees it as smooth and continuous. And if all these things are not enough of an obstacle to uniting general relativity and quantum theory, nature's non-gravitational forces operate in space-time whereas gravity *is* space-time. 'This difference may not be significant,' says Tong. 'However, gravity smells different.'

The Planck scale turns out to be significant not only because it is the scale at which gravity becomes comparable in strength to the other forces and so appears to require a quantum description. At the Planck scale, quantum theory predicts that quantum fluctuations are so big and so localised that, when energy pops into existence, it pops into existence *within its own event horizon*. This means it shrinks instantly to form a black hole. Clearly, this is nonsensical. If such events really happen, not only would space-time at the Planck scale be hidden permanently from view inside a black hole but micro-black holes would continually be born in the air all around us.

It seems that general relativity is not alone in predicting something nonsensical at the tiniest of scales: a singularity. Quantum theory also predicts something nonsensical: the spontaneous birth of black holes. The only difference is that the Planck scale, though ultra-tiny, is far short of the infinitesimal, zero-scale of the singularity. It seems that finding the deeper theory that merges general relativity and quantum theory may require fundamental modifications not just of Einstein's theory of gravity but also of quantum theory.

Even without experiments, there is a guide

The most obvious way to find the deeper theory – a quantum theory of gravity – would be to probe the ultra-tiny length scale at which Einstein's theory breaks down, at which space and time become meaningless concepts. 'In the end experiments decide about everything,' says Arkani-Hamed. 'And experiments require probing the Planck scale.'

But the ultra-tiny length of the Planck scale is synonymous with ultra-huge energy. To put things in context, the Large Hadron Collider near Geneva in Switzerland can reach collision energies of 10,000 GeV.[22] The point is that the Planck energy is 10 billion billion GeV – a *million billion* times higher than anything attainable by the LHC. Reaching such an energy with current technology would require an accelerator ring about one-tenth the diameter of the Milky Way. Perhaps somewhere out in the Universe there is an ET civilisation that has turned 10 per cent of its parent galaxy into an Ultra-Large Hadron Collider. But it seems unlikely.

The truth is there seems little realistic chance of directly probing the physics at the Planck scale. But since the entire Universe was once as small as the Planck length, there is always the possibility that the physics at that scale may have left an indelible imprint on the large-scale Universe – perhaps in the distribution of galaxies. 'We have to be looking for cosmological measurements that can get us the Planck scale,' says Arkani-Hamed.

It is possible also that the violent convulsions of space-time when the Universe was that small created powerful gravitational waves. If astronomers are clever, they might be able to see the imprint on the light of the cosmic background radiation, the 'afterglow' of the big bang fireball which is still all around us. In fact, there was a claim in March 2014 that an Antarctic-based experiment called BICEP2 had seen just such a cosmic fingerprint. Unfortunately, it turned out it had merely seen the curtain of dust that shrouds our Milky Way.[23]

It is clear that nature has put clues to the deeper theory than Einstein's way beyond our reach and that we are going to need extreme ingenuity to get even the slightest glimpse of such clues. But all is not lost. There is a powerful guide: the twin principles of relativity and quantum theory.

The undiscovered country

*The struggle to find a deeper theory than
Einstein's theory of gravity that will tell us
why there is a Universe and where it came from*

Due to the inner-atomic movement of electrons, atoms
would have to radiate not only electromagnetic but also
gravitational energy, if only in tiny amounts. As this is hardly
true in Nature, it appears that quantum theory would have
to modify not only Maxwellian electrodynamics, but also
the new theory of gravitation.

Albert Einstein[1]

There is a theory which states that if ever anyone discovers
exactly what the Universe is for and why it is here, it will instantly disappear and be replaced by something even more
bizarre and inexplicable. There is another theory which
states that this has already happened.

Douglas Adams[2]

You have climbed a lofty mountain. Getting to the summit has
taken every last ounce of energy and ingenuity, and you are exhausted but euphoric. As you pause to catch your breath, you
look up at the next mountain in the chain. And gasp. It is not
twice as high as the mountain you have just climbed, or five times
as high or even ten times as high. No. It is an impossible million
billion times as high.

This is the position that physicists find themselves in at the
beginning of the twenty-first century. They have used all the

knowhow and ingenuity of their science and technology to build the Large Hadron Collider near Geneva. It has bagged them the fabled Higgs particle, the quantum of the Higgs field, which endows all other particles with mass, and they are rightly euphoric at their spectacular success. But now, before them, stands the next great challenge: the Planck scale at which space and time and gravity likely emerge from something more fundamental and nature keeps the ultimate secret of the origin of the Universe. It is a million billion times more energetic than anything the LHC can reach and it is enough to make a grown physicist weep.

The impossibility of scaling the height of the Planck energy has led some commentators to pronounce gloomily the end of physics or, alternatively, to claim that fundamental physics has morphed into science fantasy and theorists are now free to publish any mad theories they can dream up, safe in the knowledge that no conceivable experiment will ever catch them out and prove them wrong.

Nothing could be further from the truth. 'The idea that we cannot tell that a theory is correct until we do an experiment is false,' says Nima Arkani-Hamed.

There are two physics principles that we know are true to the extent that they predict to a jaw-dropping degree of precision exactly what we see in the world that is accessible to our observations and experiments. The first is special relativity and the second is quantum theory. Physicists are not, it turns out, free to invent any old theory they like. Far from it. Their theory must be consistent with both special relativity and quantum theory. In fact, so ridiculously tight is this constraint on reality that the overwhelming majority of theories that physicists come up with are instantly ruled out. 'This is why it is so hard to find a deeper, more fundamental theory,' says Arkani-Hamed.

'Although a thousand theoretical flowers have bloomed, they still lack a firm basis in physical principles,' says historian of science Gennady Gorelik of Boston University. 'Never before in physics have so many people worked for so long with so little tangible success.'[3]

'The construction of a space-time geometry that could result not only in laws of gravitation and electromagnetism but also in quantum laws is the greatest task ever to confront physics,' said Matvey Bronstein, the man who pioneered the whole subject of quantum gravity in the 1930s.[4]

To emphasise how extraordinarily restrictive is the strait-jacket imposed on physics by special relativity and quantum theory, imagine there is a very competent physicist who knows nothing about the world and who is locked in a windowless room with two blackboards. (Do not worry yourself about how he got to be a very competent physicist while knowing nothing about the world – this is not a realistic story!) On the first blackboard are written the principles of special relativity and quantum theory. The second blackboard is largely blank except for the simple instruction: 'Deduce the consequences of other blackboard.'

For a while, the physicist contemplates the intimidating emptiness of the second blackboard. Then he picks up a piece of chalk and begins scribbling furiously. What does he write down? What does he deduce about the world?

Deducing the Universe

The first thing the physicist realises is that special relativity and quantum theory have consequences for quantum spin. As mentioned before, spin, like everything else in the microscopic world, comes in discrete chunks. The fundamental unit is ½ of a certain quantity (the quantity being $h/2\pi$).[5]

It might appear that a subatomic particle can possess any multiple of the basic unit – for instance, 19/2 or 27 or 801. But our physicist quickly discovers that nature is far more restricted and must choose its spins from an extremely reduced palette. Out of an infinity of conceivable spins, only a mere five are compatible with the twin constraints of special relativity and quantum theory: 0, ½, 1, ³⁄₂ and 2.

The spin of a particle determines how it interacts with other particles and so the phenomena for which it is responsible. Our physicist decides to consider particles of each spin, one at a time, and to write down on the empty blackboard everything he can deduce about them.

He first discovers that quantum theory requires that particles with 'half-integer' spin obey the Pauli Exclusion Principle. This endows them with a strong tendency to avoid each other.[6] The need for each such particle to have a lot of elbow room means that, when large numbers of them come together, they form spread-out, extended objects.

In fact, particles with spin ½ – which are known as 'quarks' and 'leptons' – are the fundamental building blocks of matter. A common lepton, which shares the antisocial nature of all half-integer spin compatriots, is the electron. 'It is the fact that electrons cannot get on top of each other that makes tables and everything else solid,' said Richard Feynman.

Next, our physicist considers particles of spin 1. He realises that they can be exchanged between the building blocks of matter and that this exchange gives rise to forces. There are three possibilities which lead to three distinct fundamental forces of nature.

In fact, the three 'interactions' have been given the names the electromagnetic force, the weak nuclear force and the strong nuclear force. The strong force binds triplets of quarks into protons and neutrons, and confines them in a 'nucleus'. But it has no dominion over electrons. Instead, they are bound to a nucleus by the electromagnetic force to create an atom.

Our incarcerated physicist locked in a room deduces not only the existence of ninety-two types of naturally occurring atom – from hydrogen, the lightest, to uranium, the heaviest – but the existence of a dizzying array of chemical compounds that arise from all the myriad ways in which the basic atomic building blocks may be combined.[7]

So much for particles with spin ½ and spin 1; our physicist now considers spin 0. He immediately realises that a particle with spin 0 is the 'quantum' of a 'field' which permeates all of space and resists the passage of other particles. By doing this, it endows them with inertia, or mass.

In fact, such a particle exists in the guise of the Higgs particle.

Its discovery was triumphantly announced to the world by physicists at the Large Hadron Collider in July 2012.

Next, our physicist considers spin 2. He realises that a particle with spin 2 has the property that it interacts with every other particle, giving rise to a 'universal force'. It takes a bit of calculation but he is able to show that an inevitable consequence of the existence of a spin 2 particle is the general theory of relativity.[8] This shows that special relativity is in some sense more fundamental than general relativity. How else could it be a consequence of special relativity (combined, of course, with quantum theory)?

Studying general relativity, our physicist recognises the existence of a long-range inverse-square law of attraction, which causes large bodies to orbit other large bodies. We of course know of planets that orbit stars and galaxies that orbit other galaxies. Our physicist locked in a windowless room knew of none of these. Remarkably, he has been able to deduce the existence of the large-scale Universe.

No one has yet found a particle of spin 2. And there is good reason to believe that, if it exists, it will be extremely difficult to detect. But such a particle fits the bill for the 'graviton', the hypothetical carrier of the force of gravity.[9] Since physicists have a theory of gravity in which the force of gravity is mediated by a graviton and which spawns general relativity, in a sense they already possess a quantum theory of gravity. Unfortunately, the theory is a low-energy, large-scale version of quantum gravity not the deeper version required to shed light on the ultra high energy, ultra-small length of the Planck scale.

Next, our physicist considers the one remaining spin: $3/2$. Spin $3/2$ particles permit 'supersymmetry' in which all the half-integer spin particles (fermions) are recognised to be merely the obverse face of integer-spin particles (bosons).

As yet, we have no experimental evidence that nature uses particles of spin $3/2$. But, given that it employs all other spins in its palette, there is a strong suspicion that it does. The electron, for example, is hypothesised to have a supersymmetric twin, dubbed the 'selectron'. The super-partners of known particles are considered to be good candidates for the Universe's 'dark matter', which is known to outweigh the visible stars and galaxies by a factor of about six.[10] The reason we have not yet detected supersymmetric particles, physicists suggest, is that they are very massive and that creating them requires more energy than is currently available in collisions at the Large Hadron Collider.

Although our physicist has now considered particles of every permissible spin and deduced their behaviour, there is one more thing that he can deduce from special relativity and quantum theory. Remarkably, the two principles require that each subatomic particle must have a partner with opposite electric charge or spin. Whenever a particle is created as a quantum fluctuation of the vacuum, it must always be accompanied by its 'antiparticle'.[11] For instance, a negatively charged electron is always conjured into existence alongside a positively charged 'positron'.

The Standard Model

The full inventory of the world turns out to be the following: 12 basic building blocks – 6 quarks and 6 leptons; 12 force-carriers – the photon of the electromagnetic force, 3 'vector bosons' of the weak nuclear force and 8 'gluons' of the strong nuclear force; plus the Higgs; and, of course, all the antiparticles. Collectively, these constitute the 'Standard Model' of particle physics, the triumphant culmination of 350 years of toil by physicists. It is no exaggeration to say that the Standard Model + the general theory of relativity = the World.

What is striking about the Standard Model is that so few ingredients interacting in so few ways generate so much of what we see all around us. As Gottfried Leibniz, the seventeenth-century German mathematician, so presciently observed: 'God has chosen the most perfect world – that is, the one which is the most simple in hypotheses and the most rich in phenomena.'[12]

Remarkably, our physicist, locked in a windowless room with nothing more than a blackboard and chalk, is able to deduce the main features of the world. 'Physics is shockingly constrained by quantum theory and relativity,' says Arkani-Hamed. 'They almost make the Universe inevitable.'

Almost. The twin constraints do not determine the masses of the fundamental particles nor the total number of quarks and leptons. All normal matter is assembled from just four particles – the up-quark, down-quark, electron and electron-neutrino. (A proton in an atomic nucleus, for instance, is made of two up-quarks and a down-quark, and a neutron two down-quarks and an up-quark.) But nature has not stopped here. It has created heavier versions of the basic four particles – the strange quark, charmed quark, muon and muon-neutrino – and heavier-still versions – the bottom quark and top quark, tau and tau-neutrino. Such particles play essentially no role in the Universe today since the energy to create them existed only in the first split-second of the big bang. To paraphrase the American physicist I. I. Rabi: 'Who ordered them?'[13]

The Standard Model does not reveal why nature has triplicated its basic building blocks – or why it has given the fundamental particles the masses they have. It is a strong indication that it is not the final word on nature but merely an approximation of a deeper theory, yet to be found. But these shortcomings should not detract from the fact that the principles of special relativity and quantum theory are so tight a constraint on the possible that they determine pretty much everything about the physical world. 'What really interests me is whether God had any choice in the creation of the World,' said Einstein. The lesson of special relativity and quantum theory appears to be that He did not.

As mentioned at the outset of this chapter, some people claim that theoretical physicists are fantasists who spend their time imagining all sorts of weird and wonderful things which are so impossibly beyond the reach of experimental test that they can never be proved wrong. But the fact that special relativity and quantum theory pretty much uniquely determine the Universe around us can mean only one thing: they are largely correct. This, in turn, means they are a severe straitjacket on any deeper theory invented by physicists. So little wriggle room is there that finding theories that fit inside is extremely difficult. 'Almost everything you try fails,' says Arkani-Hamed. 'The overwhelming majority of theories that physicists can imagine are killed at birth.'

In fact, in 2017, there is only one candidate for a deeper theory that satisfies the constraints of both special relativity and quantum theory: 'string theory'.[14]

A wonderful thing is a piece of string

String theory – also known as superstring theory – arose out of attempts to understand nature's strong nuclear force. The force is not called strong for nothing. So much energy must be put into pulling apart a pair of quarks that, in the space between them, it spontaneously creates a quark–antiquark pair. Think of trying to reach a friend in a crowd as other people constantly and annoyingly insert themselves between you. This is the way it is for quarks. The strong nuclear force imprisons quarks within the protons and neutrons of atomic nuclei and makes it impossible to isolate a lone quark.[15]

The strong nuclear force is also weird in that its attraction gets stronger the further apart are two quarks. Contrast this with a familiar force such as gravity, which weakens the greater the separation of two masses; or magnetism, which weakens the greater the separation of two magnets. The reason that these forces become diluted is that they leak out in all directions.[16] But if a force is instead confined to a narrow channel between two bodies, it can strengthen the further apart they are pulled. This is the case for the force in a stretched spring or an elastic band.[17] And it is also the case for the strong nuclear force between two quarks. This behaviour was the first indication that the fundamental building blocks of the Universe, instead of being point-like subatomic particles, may actually be tiny one-dimensional strings of energy.

In the rudimentary theory pioneered by Italian physicist Gabriele Veneziano in 1968, such entities vibrate much like violin strings, and each possible vibration corresponds to a different fundamental particle.[18] 'In essence, string theory describes space and time, matter and energy, gravity and light, indeed all of God's creation . . . as music,' says writer Roy H. Williams.[19]

In the way that a rapidly vibrating violin string is more

energetic than a sluggishly vibrating violin string, a rapidly vibrating string corresponds to a subatomic particle with a lot of mass-energy, such as a top quark, while a sluggishly vibrating string corresponds to a particle with not much mass-energy, such as an electron. But, because of the complexity of the mathematics, physicists cannot yet be sure that all the possible vibrations of strings can account for all the known fundamental particles.

Strings can either be open-ended, or closed to form a loop. Whether a string is open or closed determines the way in which it interacts with other strings.

String theory automatically includes a half-integer spin (matter) partner for every integer spin (force-carrier), and vice versa. It is because the theory therefore incorporates supersymmetry that it is called 'superstring' theory and the strings are referred to as 'superstrings'. As mentioned before, none of the superpartners of the known particles has been found, though string theorists maintain they are simply too massive to have been created at the LHC.

String theory resolves a potential conflict between two powerful ideas in physics. 'Reductionism' is the idea that the phenomena of the world are the result of the interaction of a handful of fundamental building blocks. In the Standard Model, those building blocks are quarks and leptons. 'Unification' is the belief that disparate phenomena of nature are in fact facets of a single, more fundamental phenomenon. Electric and magnetic fields, for instance, are merely aspects of the 'unified' electromagnetic field.

Reductionism, taken to its logical conclusion, is expected to reveal the world to be built out of a single type of building block. But, if such a building block is truly fundamental – that is, composed of no internal parts that can be rearranged – how can it have different faces? The answer is it cannot if it is a point-like particle. It can, however, if instead it is a one-dimensional string, capable of a multitude of different vibrational modes. Thus strings neatly avoid the conflict between unification and reductionism.

Fundamental particles not only have distinct masses – which

can be mimicked by the vibration rate of a string – they also interact via fundamental forces. In 1915, Einstein showed that the 'force' of gravity is nothing more than a manifestation of the warpage of four-dimensional space-time. In the 1920s, two physicists took Einstein's idea a step further. Independently of each other, Theodor Kaluza and Oskar Klein showed that if there exists another space dimension, making space-time five-dimensional, the force of gravity *and* electromagnetism can both be consequences of the curvature of space-time. Such an extra space dimension is not obvious. But it might have gone unnoticed, claimed the two physicists, if instead of being a big dimension like north–south, east–west and up–down, it is curled up smaller than an atom.

In Kaluza and Klein's scheme, a subatomic particle, even when at rest in normal space, is whirling ceaselessly round and round in the extra dimension like a demented hamster in a wheel. In fact, momentum in the extra dimension *is* electric charge. And the reason electric charge comes in multiples of a basic chunk, or is 'quantised', is that particles behave as waves, and the only waves permitted are those with a wavelength that fits around the circumference of the extra dimension once, twice, three times and so on. Such waves necessarily have a momentum (charge) that is a multiple of the momentum (charge) of the longest permitted wave.

In the 1920s, when Kaluza and Klein proposed their idea, nature's strong nuclear force and weak nuclear force, which hold sway only in the ultra-tiny domain of the atomic nucleus, had yet to be discovered. But it is perfectly possible to mimic the behaviour of these extra forces using yet more extra space dimensions, each rolled up so small they are unnoticeable. In fact, a total of six extra space dimensions are required. The hypothetical strings of modern string theory consequently quiver in ten-dimensional space-time – nine dimensions of space and one of time.

'Einstein comes along and says, "Well space and time can warp and curve – that's what gravity is,"' says physicist and popular science writer Brian Greene of Columbia University in New York. 'And now, string theory comes along and says, "Yes, gravity, quantum mechanics, electromagnetism all together in one package, but only if the Universe has more dimensions than the ones that we see."[20]

'At first people didn't like extra dimensions,' says string theorist Edward Witten of the Institute for Advanced Study in Princeton, 'but they've got a big benefit. The ability of string theory to describe all the elementary particles and their forces along with gravity depends on using the extra dimensions.'

The pros and cons of string theory

A theory which maintains that space-time is ten-dimensional is seriously in conflict with the fact we appear to live in a three-dimensional reality (four-dimensional if you count time). But the theory has other problems as well. For a start, the strings are postulated to be mind-bogglingly tiny – equivalent to the Planck length, which is 10^{-35} metres, or a million billion times smaller than a hydrogen atom. Consequently, even the most violent particle collisions at the LHC have a million billion times too little energy to directly probe the world of strings. And because the strings exist at an energy scale and size scale so far removed from the everyday realm, they produce no noticeable imprint on the familiar world. So not only are strings inaccessible to terrestrial experiments, their existence leads to no testable predictions. 'It is wonderful that both the Standard Model and general relativity come out of string theory,' says David Tong. 'But, really, what physicists would like to come out is something unpredictable.'

String theory also requires nature to use supersymmetry. As the LHC explores ever higher energy regimes, there are fewer and fewer places for supersymmetric particles to hide. If they do not come to light soon, the theory could be dead in the water. String theory is a beautiful mathematical structure, something which even its critics accept. But there are many equally beautiful ideas which nature, in its wisdom, has chosen not to implement.

Yet another problem with string theory is that there are a huge number of ways in which the extra dimensions can be intertwined. According to some estimates, this leads to at least 10^{500} distinct 'string vacua', in each of which the numbers and masses of fundamental particles are different, and even the number of fundamental forces and their respective strengths. Physicists had imagined that, because special relativity and quantum theory are so very hard to unite, any framework that united them would be unique, predicting the observed properties of the fundamental particles and fundamental forces. 'This idea was wrong,' says Arkani-Hamed.

Instead, physicists have found a mind-bogglingly large number of 'solutions' of string theory, all of which are compatible with special relativity and quantum theory. In fact, string theory has been called 'a bunch of solutions in search of a theory'. This is by no means unprecedented in physics. For instance, there are an infinite number of possible electromagnetic waves, each with a distinctly different wavelength. All are solutions of Maxwell's equations of electromagnetism. Then there are hydrogen atoms and tables and you, at this moment, reading these words. All are solutions of the Schrödinger equation.

The big question is: what is the underlying theory to which the 10^{500} string vacua are the solutions?

At one time, string theorists were exploring five distinctly different variations of string theory, which went by the name of Type I, Type IIa, Type IIb, Heterotic O (32) and Heterotic $E_8 \times E_8$. But, in the mid-1990s, Paul Townsend of the University of Cambridge and Chris Hull of Queen Mary College, London, showed that all five are merely different ways in which supersymmetry is realised – different versions of a single eleven-dimensional theory. 'M-Theory' was named by Edward Witten but he has never stated what the 'M' stands for. 'Eleven-dimensional M-theory is the umbrella theory,' says David Berman of Queen Mary University in London.

The 10^{500} string solutions are solutions of M-Theory. Collectively, they look a lot like an ensemble of universes, or a 'multiverse', except that they are probably all connected to each other. Science-fiction writer Arthur C. Clarke could have been writing about string vacua when he wrote: 'Many and strange are the universes that drift like bubbles in the foam upon the River of Time.'[21]

Physicists would have preferred a theory that predicts the precise properties of the fundamental particles and fundamental forces. Instead, they must answer the question: why are we in the string vacuum we are in and not any of the 1-followed-by-500-zeroes others? 'We don't know yet,' says Arkani-Hamed.

One way would be to count how many universes there are with each possible value of the electron mass, with each possible strength of the electromagnetic force, and so on. The most common universes should be ones whose subatomic particles have masses very close to ours and whose fundamental forces have strengths very close to ours. If we discover that we live in a very special and unusual Universe then that would be inexplicable and string theory would be dealt a serious blow. 'The problem is no one can think how to count up universes,' says Arkani-Hamed.

Berman is not unduly worried by this. 'It is too early to be discouraged yet in our exploration of the mathematical structure of string theory,' he says. 'We are still a long way from real physics.'

Despite all the difficulties with string theory, however, it boasts a number of compelling features, which keep a large global community of physicists not only interested but positively enthusiastic about it. Most importantly, the theory contains a vibrating loop of string of spin 2. A particle of spin 2, as mentioned before, is the recipe for the graviton, the carrier of the gravitational force. Not only that but an unavoidable consequence of the existence of a spin 2 particle is the general theory of relativity. As already mentioned, the Holy Grail of physics is the uniting of quantum theory and Einstein's theory of gravity. It is fantastically appealing that string theory is a quantum theory that automatically incorporates the general theory of relativity.

But, to Berman, it is its richness that makes string theory compelling, not just the fact it contains a quantum theory of gravity. He compares string theory with Newton's theory of gravity. 'It explained not just one thing but many things – the motion of planets, the ocean tides, the precession of the equinoxes, and so on – and it gave physicists things to work on effectively for ever,' says Berman. 'Similarly, we are far from finished exploring string theory. It can run and run.'

Until 1985, string theory languished in a backwater of physics, pursued by only a handful of aficionados who were convinced of its promise. Everything changed when John Schwarz of the California Institute of Technology in Pasadena and Mike Green of Queen Mary College in London made a breakthrough.

Physics contains many symmetries – aspects of a situation that remain the same when something else is changed, in much the same way that a square continues to look like a square if rotated by a quarter of a turn, half a turn, and so on. In 1918, the German mathematician Emmy Nöether made the remarkable discovery that symmetries underpin many of the great laws of physics. Take the law of conservation of energy, which says that energy cannot be created or destroyed, merely morphed from one form into another. This is a consequence of 'time-translational symmetry', the fact that, if a particular experiment is done next week or next month or next year, the outcome, all things being equal, will be exactly the same.

Nöether's recognition that symmetry underlies the laws of physics is one of the most powerful ideas in modern physics and the reason why physicists at the LHC hunt for symmetries as indicators of new fundamental laws. Many theories, when quantised do not preserve classical symmetries such as the key 'Lorentz symmetry' of Einstein's special theory of relativity. Schwarz and Green discovered that string theory does. In the jargon, it is 'anomaly free'. 'All the symmetries of classical physics automatically apply,' says Berman. 'Miraculously, string theory is compatible with everything we already know to be true.'

Schwarz and Green's discovery triggered the 'first string revolution', which saw string theory transformed from a niche research area into a mainstream field of enquiry. The 'second string revolution' was the realisation that string theories are versions of M-Theory, and that, ironically, the most important things in string theory are not strings.

Brane power

The three-dimensional everyday world contains not only one-dimensional objects such as lengths of cotton but two-dimensional objects such as table-tops and three-dimensional objects such as trees and people. By analogy, the M-Theory Universe of ten space dimensions may contain not only one-dimensional strings but objects with two dimensions, three dimensions, and so on . . . all the way up to ten dimensions. Collectively, physicists refer to these multi-dimensional entities as 'branes', more colourfully christened 'p-branes' by Townsend, where the p refers to the number of space dimensions. A string, in this terminology, is a 1-brane.

In M-Theory, it turns out that branes are not only possible, but required. And this proliferation of multi-dimensional objects means the theory is unlikely to be one in which strings play a fundamental role. It appears instead that a large ensemble of objects is needed to unite special relativity and quantum theory. 'In some domains of the theory, particle phenomena will manifest themselves, in others string phenomena, in still others the phenomena of 2-branes, 3-branes, and so on,' explains Arkani-Hamed.

In the brane picture, the large-scale Universe is a three-dimensional island, or 3-brane, floating in a space of ten dimensions. In this scenario strings have two possibilities open to them. One end of a string can be anchored to the 3-brane, waving free like kelp attached to the seabed of the Sargasso Sea. Or a string can form a loop which is not attached to the 3-brane. The familiar fundamental particles of the Standard Model are of the former type and so confined to the 3-brane. The graviton alone is a string loop which is free to move off-brane and explore the ten-dimensional 'bulk'.

This suggests an intuitive explanation of one of the biggest puzzles in physics: why the force of gravity is weaker than the other fundamental forces of nature by such an extraordinarily large factor. As pointed out before, the force of gravity between a proton and electron in a hydrogen atom is 10,000 billion billion

billion billion times weaker than the electromagnetic force between them. In 1999, Lisa Randall of Harvard University and Raman Sundrum of the University of Maryland in College Park discovered that extra dimensions need not be rolled up far smaller than an atom. If they were warped in a particular way, they could be as large as the Universe but have gone entirely unnoticed.[22]

In the Randall-Sundrum scenario, the reason the force-carriers of the non-gravitational forces such as the electromagnetic force are relatively strong is that they are confined to our 3-brane. Gravitons, on the other hand, leak out into the full ten-dimensional bulk, and so their effect is diluted.

Although this is an attractive intuitive explanation for the weakness of gravity, there is no evidence yet of large space dimensions that are hidden from our view. String theory remains big on possible explanations of phenomena in our Universe but small on actual explanations that are capable of spawning precise and testable predictions.

If our Universe is indeed a three-dimensional island floating in a ten-dimensional space-time, then an obvious thing to wonder is: Is it the only such island? And if it is not the only island, could our 3-brane collide with another 3-brane? This is actually the basis of a novel explanation for the big bang proposed by a team led by physicist Neil Turok, director of the Perimeter Institute in Waterloo, Canada.

In the scheme, two entirely empty 3-branes approach each other along a fifth dimension (the fourth being time). Think of them as two slices of bread coming together, flat-side to flat-side. The two 3-branes pass right through each other. But they have enormous energy of motion in the fifth dimension and, at the moment they kiss, that energy has to go somewhere. Where it goes is into creating the mass-energy of subatomic particles on the branes and into heating them to a blisteringly high temperature. In short, it creates a hot big bang.

In Turok's scheme, on each brane the fireball expands and cools, galaxies congeal out of the debris and fly apart, eventually diluting the matter to such an extent that each brane is essentially empty again. The vacuum in the fifth dimension acts like a spring, which eventually pulls the branes back together again. They collide and repeat the cycle. Again. And again . . . Our big bang is just one in a long line of bangs, stretching back into the past and forward into the future.

The 'cyclic Universe' is potentially distinguishable from the standard cosmological scenario in which the Universe in its first split-second undergoes a phenomenally violent, exponential expansion known as 'inflation'. 'If the Universe sprung into existence and then expanded exponentially, you get gravitational waves travelling through space-time,' says Turok. 'These would fill the Universe, a pattern of echoes of the inflation itself.' The cyclic Universe, on the other hand, lacks the chaotic violence necessary to shake the hell out of space-time and so predicts no such gravitational waves from the early Universe.

The cyclic Universe is a speculative idea. String theory itself is not a fully fleshed-out theory. It may be only a small part of the deeper theory that will explain the origin of space and time and the Universe. Or it may be a red herring. But string theorists feel encouraged that they are on the right track. One reason is of course that it is the only game in town – despite immense effort, no one has found another 'theory of everything' capable of unifying all the fundamental forces. But another reason for the optimism of string theorists is that their theory potentially resolves a paradox involving the most mysterious objects in the Universe: black holes.

Black holes

It is at the very heart of black holes – where Einstein's theory of gravity predicts that matter is crushed to infinite density – that known physics well and truly breaks down. But the singularity is not the only location in a black hole that challenges our understanding of reality.

The 'event horizon', as mentioned before, is an imaginary membrane surrounding the singularity that marks the point of no return for in-falling light and matter. When people talk of the size of a black hole, they are implicitly referring to the size of the horizon.

In 1974, Stephen Hawking shocked the world of physics by announcing that black holes are not actually black. He came to this conclusion after considering quantum processes in the vicinity of a black hole. Remember that the Heisenberg Uncertainty Principle permits particle–antiparticle pairs to be conjured into existence out of the vacuum. Such 'virtual' particles have a fleeting existence, annihilating each other and popping back out of existence in a lot less than the blink of an eye. But Hawking realised that, just outside a black hole's event horizon, something profoundly different can and does happen.

One member of a newly created particle–antiparticle pair can race outwards, escaping for ever the gravity of the black hole, while the other falls through the horizon into the interior of the black hole. Once trapped inside, it can never re-emerge to annihilate with its birth partner. The escaping particle is elevated from the status of a short-lived virtual particle to a long-lived real particle.

Hawking realised that processes like this are occurring continually all around the horizon of a black hole. They cause it to glow with 'Hawking radiation' as particles stream outwards.

A defining characteristic of a black hole is, of course, that nothing inside can ever come out. Hawking radiation does not come out of a black hole since it is never in any sense inside. Instead, it is born in the vacuum just beyond the edge of the event horizon.

The energy to make Hawking radiation real has to come from somewhere. And the only source is the gravitational energy of the black hole itself. As particles continually stream away into space, the gravitational field of the black hole weakens, causing it to gradually shrink, or 'evaporate'.

The smaller a black hole the stronger its Hawking radiation.[23] For stellar-mass black holes and the supermassive black holes found at the heart of most galaxies, the particle sleet is so utterly negligible that their life expectancy far exceeds the current age of the Universe. But, as a black hole shrinks, its Hawking radiation gets ever stronger. For a tiny black hole – and every black hole reaches a tiny size of course before it finally vanishes – the Hawking radiation is blindingly bright. Black holes, when they end their lives, go out with a bang not a whimper.

Anything that glows, by definition, has a temperature. And this is the case for a black hole shining with Hawking radiation. At first sight this seems bizarre because a black hole is nothing more than a bottomless well in space-time and so contains no obvious source of warmth. But a black hole is hot not by virtue of any intrinsic property but by virtue of extrinsic quantum processes going on in the surrounding vacuum.

The tendency of Hawking radiation to cause a black hole to evaporate and eventually disappear creates a serious scientific paradox. It is a fundamental law of physics that information cannot be created or destroyed. Take the Moon. The fact that its location tomorrow can be predicted from its location today by the application of Newton's laws implies that information about its location tomorrow is contained within its location today. So, as the Moon travels across the sky, information is neither gained nor lost but 'conserved'. In the evaporation of a black hole, however, information *is* lost.

The precursor of a stellar-mass black hole is of course a star. A vast amount of information is needed to precisely define such a celestial body. The description must include, for instance, the type, location and velocity of every one of the star's atoms. But, when a black hole has evaporated completely via Hawking radiation, there is literally nothing left. Where does all the information go? This, in a nutshell, is the 'black hole information paradox'.

So baffling is the paradox that, for many years, Hawking himself actually entertained the idea that black holes did indeed violate one of the most cherished principles of physics. 'I used to think information was destroyed in a black hole,' he said. 'It was my biggest blunder, or at least biggest blunder in science.'[24]

The obvious culprit for a repository of the missing information is the Hawking radiation. Perhaps it somehow spirits away knowledge of the star that spawned the black hole to the four corners of the Universe? But the Hawking radiation – which technically has the spectrum of a 'black body' – is characterised solely by its temperature.[25] All it carries away from the black hole is this one ultra-trivial piece of information.

A clue to the resolution of the 'black hole information paradox' came from Israeli physicist Jacob Bekenstein. In 1972, he discovered something unexpected about the event horizon: its 'surface area' is related to the black hole's 'entropy'.[26]

Entropy is a concept that emerges from the theory of heat. 'Entropy always increases' is a statement of the 'second law of thermodynamics', one of the most important principles in science, which explains why castles crumble, eggs break and people grow old. Bekenstein's discovery was the first indication of a connection between black holes and heat, pre-dating Hawking's discovery that such bodies glow with heat radiation. Uniquely in black holes, three of the great theories of physics collide: Einstein's theory of gravity, quantum theory and 'thermodynamics', the theory of heat. This is why gaining an understanding of them is of such key importance in the quest to mesh together quantum theory and the general theory of relativity.

Entropy is intimately connected to information. It is a measure of our lack of information, or ignorance of the state, of a system. More specifically, it is a measure of the system's microscopic disorder and is defined as the 'number of microstates that correspond to a particular macrostate'. In the case of a brick, for instance, it is the number of different ways the atoms in the brick can be arranged and still leave the brick looking like a brick. The fact that the horizon of a black hole has an entropy can mean only that it is not smooth and featureless, as general relativity maintains, but has some kind of microscopic structure.

In 1993, Dutch Nobel Prize-winner Gerard t'Hooft of the University of Utrecht suggested that the horizon of a black hole, far from being smooth and featureless, is rough and irregular on the microscopic scale. And it is in the lumps and bumps of its Lilliputian landscape that is stored the information that describes the star that gave birth to the black hole. It is as if the horizon is a super-dense DVD, with every region a Planck length on a side – that is, of area about 10^{-70} square metres – containing the equivalent of a binary '0' or '1' of information. 'A black hole really is an object with very rich structure, just like Earth has a rich structure of mountains, valleys, oceans, and so forth,' says Kip Thorne of the California Institute of Technology in Pasadena.

Shortly after t'Hooft's proposal that the missing information in a black hole might be encoded in its event horizon, Leonard Susskind of Stanford University showed how it might be implemented in string theory. Think of the event horizon of a black hole as a squirming mass of vibrating strings. Using this picture, in 1997, Andrew Strominger of the University of California at Santa Barbara and Cumrun Vafa of Harvard University were able to predict the exact black hole entropy calculated by Bekenstein.[27]

Since Hawking radiation is born in the vacuum just a hair's breadth above a black hole's event horizon, it stands to reason that it is influenced by the microscopic undulations of that membrane. Those undulations 'modulate' it in much the same way that pop music modulates the 'carrier wave' of a radio station. In this way the information that described the precursor star is carried out into the Universe imprinted indelibly on the Hawking radiation. No information is lost, after all. And one of the most precious laws of physics is left intact.

This proposal for averting the black hole information paradox remains speculative. We still lack the deeper theory that will mesh together Einstein's theory of gravity and quantum theory. But, if correct, it implies something extraordinary. The information to completely describe a star – a 3D body – is perfectly preserved on the horizon of a black hole – a 2D surface. This makes the horizon similar to the holographic image on a credit card. Say a frog carried around with it a hologram of its previous incarnation as a tadpole. Well, a black hole carries around with it a hologram of its previous incarnation as a star.

This might be nothing more than a weird curiosity if it applied only to esoteric objects such as black holes. But t'Hooft and Susskind suggested that the holographic idea might have

implications not just for black holes. It may say something profound about the whole Universe.

The holographic Universe

The Universe, in common with a black hole, is surrounded by a
horizon. The cosmic 'light horizon' is not the edge of the Universe – which plausibly may go on forever – but defines the edge
of the 'observable Universe'. Within the horizon are all the stars
and galaxies whose light has had time to reach us since the birth
of the Universe 13.82 billion years ago. Outside the horizon are
all the stars and galaxies whose light has had insufficient time.
Their light is still on its way.[28]

T'Hooft and Susskind reasoned that just as the information
that describes a 3D star is inscribed on the 2D horizon of a black
hole, the information that describes the 3D Universe might be
written as a 2D hologram on the horizon of the Universe. The
idea is open to several possible interpretations. One is that the
Universe, for some unknown reason, can be completely specified
using one fewer large dimensions than expected. This is bizarre
enough. Another, wilder, interpretation is that we actually live on
the surface of the horizon, believing that we occupy its interior.
Yet another, even wilder, interpretation is that our 3D Universe
is literally a projection of a 2D hologram residing on the horizon
that surrounds it, in which case, you and I and everyone else is
actually a hologram!

Reasoning by analogy in such a manner is hardly rigorous
physics. And it is a big leap to extrapolate from the properties
of black holes to the properties of the entire Universe. But, in
1998, the Argentine physicist Juan Maldacena published a paper
which not only shored up the idea that we live in a 'holographic
Universe' but set the world of physics alight.

'Conformal field theories' are a class of theories that are compatible both with quantum theory and special relativity (the
Standard Model is one such theory). Maldacena pictured a 5D
Universe, dubbed 'the bulk', filled with fundamental particles
that dance to the tune of Einstein's theory of gravity. He then

pictured this Universe's 4D boundary, which encloses it much as the 2D surface of a balloon encloses a 3D volume of air. It contained fundamental particles dancing to the tune of a conformal field theory.[29]

Maldacena's 'miraculous' discovery was that the equations on the boundary contain the same information and describe the same physics as the more complex equations of the bulk. In other words, the effects of gravity in the interior are mathematically equivalent to quantum field theory on its boundary. 'The duality between a quantum description and a gravitational description appears to reveal a deep and surprising connection between quantum theory and Einstein's theory of gravity,' says Berman. 'Although they appear utterly different, they may in fact be different sides of the same coin.'

'Quantum theory and relativity appear to fight each other,' says Arkani-Hamed. 'But, behind the scenes, they actually help each other.'

So significant has Maldacena's paper been deemed by the physics community that it has received almost 10,000 citations in other scientific papers and is widely regarded as a milestone in modern physics. Some physicists believe that the discovery of a link between gravity and quantum theory could turn out to be as important as the discovery by Maxwell in the nineteenth century that a single theory connects electricity, magnetism and light.

Berman cautions that Maldacena's result applies only in a simplified, or 'toy', model of the Universe known as 'Anti de-Sitter', or AdS. Among other things its space does not expand like normal space. But the hope of physicists is that the result also applies in the real Universe. But nobody has yet been able to prove it.

What is space?

The key question posed by Maldacena's discovery is: how does a quantum field on the boundary produce gravity in the bulk? In an attempt to answer this, in 2015 Mark Van Raamsdonk of the University of British Columbia in Vancouver imagined an even simpler model than Maldacena's. It was an empty bulk Universe. This corresponds to a single quantum field on the boundary. Like all quantum fields, it was tied together with entanglement – the instantaneous influence Einstein called 'spooky action at a distance'.[30]

Using mathematical tools developed by others, Van Raamsdonk was able gradually to remove the entanglement on the boundary. As he did so, he saw that the space-time of his Universe steadily elongated, pulling apart as if made of toffee. Gradually, the very structure of space-time began to disintegrate. Eventually, when he had reduced the entanglement to zero, the space-time shattered into fragments like toffee stretched too far.

Van Raamsdonk concluded that the long-distance connections which quantum entanglement bestows on space-time are essential to knit it together into a smooth whole. 'Space-time is just a geometrical picture of how stuff in the quantum system is entangled,' he says.[31]

If Van Raamsdonk is correct, space-time might actually emerge out of 'quantum information'. But the result applies only to a simplified toy model and no one has been able to prove that it applies in the real Universe. Nevertheless, the idea that entanglement is essential for space-time to exist appears to be supported by another line of reasoning.

In 2013, Maldacena and Susskind drew people's attention to two papers by Einstein, both coincidentally published in 1935. On the surface, they address subjects that could not be more different. But Maldacena and Susskind conjectured that they might actually be intimately related.

In the first paper, Einstein, Boris Podolsky and Nathan Rosen highlighted the quantum phenomenon of entanglement and pointed out (incorrectly) that such 'spooky action at a distance' is so ridiculous that it can mean only that quantum theory is flawed and incomplete.[32] In the second paper, Einstein and Rosen showed that short-cuts through space-time are permitted to exist by the general theory of relativity.[33] Today, we know them as 'wormholes', a term coined by the American physicist John

Wheeler, who also coined the term 'black hole'. Just as a worm-hole through the centre of an apple enables a worm to take a shortcut to the far side rather than crawl around the surface, a space-time wormhole would enable a space traveller to take a shortcut across the Universe. After entering one mouth, they might need to crawl only a few metres before being spat out of the other mouth on the far side of the Galaxy.

According to Maldacena and Susskind, the connection that physicists call a wormhole is equivalent to quantum en-tanglement. In other words, if two particles are connected by entanglement, they are effectively joined by a submicroscopic wormhole. Remarkably, wormholes in space-time and quantum entanglement may merely be different ways of describing the same underlying reality.

If entanglement occurs because of the existence of microscopic wormholes in space-time – and such wormholes are essential for the very existence of space-time – then reducing entanglement would be expected to damage the fabric of space-time, just as found by Van Raamsdonk. The answer to the question 'What is space made of?' may well be quantum entanglement/wormholes. Take your pick. According to Maldacena and Susskind they are the same basic phenomenon.

Dazzled by dualities

Maldacena's demonstration that 'quantum field theory' on the horizon of a 5D Universe manifests itself as general relativity in the space inside the horizon is an example of the existence of superficially very different depictions of the same physical situation. The existence of such 'dualities' often enables a problem that is impossible to solve from one point of view to be easily soluble from another point of view. And string theory, it turns out, is awash with dualities.

A very typical duality of string theory is that the physics at ultra-small scales looks exactly like the physics at ultra-large scales. The origin of 'T-duality' is in the fact that strings can both move around or wrap around an extra space dimension, so momentum can be exchanged with winding. This takes the physics from the small to the large and vice versa.

A key consequence of this particular duality is that, on the smallest scales, the parameters of physics such as the strength of gravity do not skyrocket to infinity as predicted by Einstein's theory of gravity. Instead, they remain well-behaved as they would on the largest scales. Intuitively, this makes sense since strings have a finite size. Since they cannot be squeezed into zero volume, this neatly avoids the catastrophic singularity that general relativity predicts for the beginning of the Universe.

Dualities are by no means unique to string theory. They are also found in other fields of physics such as quantum theory, which is of course famous for its wave–particle duality. In truth, though, the separate wave and particle ways of looking at the microscopic building blocks of matter were the currency of discussion only while quantum theory was in the process of development – and continue to be currency only in popular-science books like this one! With the discovery of a self-consistent framework of quantum theory in the mid-1920s, wave–particle duality went away. The quantum machinery of Schrödinger and Heisenberg manipulates mathematical entities such as 'wave functions' that are neither wave nor particle but something for which we have no word in our vocabulary nor any analogue in the everyday world.

In much the same way that wave–particle duality was an indication that an adequate theory of the quantum still eluded physicists, the dualities in string theory indicate that string theory is not complete. 'We are not quite there yet,' says Berman. 'The true, deeper theory will have no dualities.'

But how do we find the deeper theory?

Finding Neverland

Arkani-Hamed thinks that, in searching for a deeper, more fundamental, more true theory of physics there are several possible strategies that physicists might employ. The obvious one is to make a list of all the assumptions that have gone into obtaining the current picture, then cross them out one at a time in the hope that, gradually, the current best theory will morph into the much sought-after deeper theory. 'Historically, however, this strategy has never worked,' says Arkani-Hamed.

For reasons nobody knows, theories of physics are like Russian nesting dolls. Inside each perfect doll is another perfect doll. Similarly, behind each perfect and self-contained theory of the world, physicists find a deeper theory that is also perfect and self-contained. There is no gentle morphing of one theory into a deeper theory. Nature is simply not like that. 'The laws of physics at one level are perfect,' says Arkani-Hamed. 'And, at the deeper level, they change into laws that are even more perfect.' And the only way to get from one to the other is to make a heart-stopping leap in the dark. As Newton said: 'No great discovery was ever made without a bold guess.'

Classical physics and quantum theory are a prime example of nature's nesting-doll tendency. In the late nineteenth century, classical physics appeared perfect and self-contained. It had an apparently minor shortcoming in the guise of the ultraviolet catastrophe, considered important by Planck, and to a greater extent, by Einstein. But the deeper theory that fixed the ultraviolet catastrophe did not in any sense grow out of classical physics. The discovery of quantum theory instead essentially involved plucking out of thin air new principles and new equations, such as the Schrödinger equation, which were utterly incompatible with classical physics and in no sense could be deduced from it.

The way in which the laws of physics appear not to morph smoothly from one level to a deeper level but rather to change abruptly, even seismically, leaves physicists with only one option, says Arkani-Hamed. 'Hold onto the physics we know for as long as possible, then jump!'

The physics we know is special relativity and quantum theory, and the only framework we have so far found that unites them is string theory. Arkani-Hamed advocates pushing such physics to breaking point. Then, like a man who has reached a cliff edge in the dark, taking a leap into the void in the hope of parachuting down onto a new island of physics off the coast. 'Physics makes progress in a very discontinuous way,' says Arkani-Hamed. 'It is very important to be in the vicinity of the answer and so to leap from the right spot.'

The deeper theory will supplant Einstein's theory of gravity, which of course breaks down at the singularities at the heart of black holes and at the beginning of time. 'But it may also require an extension of quantum theory,' says Arkani-Hamed.

'Most theories signal their own demise – the theory of electromagnetism with the ultraviolet catastrophe, the general theory of relativity with its singularities – but quantum theory doesn't seem to,' says Berman. 'With quantum theory we have seen something very deep.'

Although quantum theory is currently 'fit for purpose' – in that it predicts perfectly the outcome of all experiments – it assumes the existence of a universal clock marking time. 'If the notion of time breaks down close to singularities, however, it is not clear that quantum theory can provide any guide for us,' says Arkani-Hamed. 'It's only in the domain of cosmology – which deals with the origin, evolution and fate of the Universe – that quantum theory is potentially in trouble.'

'The deeper theory will be neither general relativity nor quantum theory but a third theory,' says Lee Smolin of the Perimeter Institute in Waterloo, Canada.

The problem in taking the next step is that all the fragmentary theories and insights gained from toy models of reality need to be put together. But nobody knows which ones are correct. And they may all be wrong. 'String theory is part of the deeper theory,' says Arkani-Hamed. 'But it may not even be a central part.'

Up is the new down

When Arkani-Hamed proposes finding the deeper theory by stretching known physics to breaking point then making a leap into the unknown, he is implicitly assuming that we are in possession of all the observational data necessary to answer those big questions. Currently, we know of twelve building blocks of matter – six quarks and six leptons – and of four fundamental forces. But the familiar atomic matter that makes up the stars and galaxies and you and me is outweighed by a factor of about six by the mysterious dark matter. 'Dark matter may be absolutely critical,' says Arkani-Hamed. 'There could be a property of the Universe that is a game-changer – that shows us that string theory is wrong.'

It is not possible, for instance, to rule out the existence of dark particles and dark forces that might change profoundly our understanding of physics. 'There are more things in heaven and earth, Horatio, than are dreamt of in your philosophy,' warned Shakespeare's Hamlet.

It is a remarkable fact that only about 4.9 per cent of the mass-energy of the Universe is normal matter – the stuff of the Standard Model – and, of that, we have so far spotted only about half with our telescopes. The remainder is suspected to be hydrogen gas floating between the galaxies, which is either too cold or too hot to give out detectable light.[34] By comparison, the dark matter accounts for about 26.8 per cent of the mass-energy of the Universe and the 'dark energy' about 68.3 per cent.

As mentioned before, the dark energy – despite being the major mass component of the Universe – was discovered only in 1998. It is invisible, fills all of space and has repulsive gravity. In fact, its repulsive gravity is what is speeding up the expansion of the Universe and what led to its discovery.[35]

If schools are still teaching children that gravity is an attractive force, they are behind the times. More than two-thirds of the stuff in the Universe – the stuff causing its expansion to accelerate – has repulsive gravity. 'We know there is gravity because apples fall from trees. We can observe gravity in daily life,' says dark energy researcher Adam Riess of Johns Hopkins University in Baltimore. 'But if we could throw an apple to the edge of the Universe, we would observe it accelerating.'

Dark energy probably does not have the potential to throw a spanner in the works as damaging as does dark matter, since general relativity and quantum theory predict the existence of vacuum energy – though nobody knows how the two dovetail together.[36]

So much for missing observational data about our Universe – are we also missing a big idea? 'Our framework is spectacularly right in many ways,' says Arkani-Hamed. 'But it is also obvious that something big is wrong. The next step will require some revolutionary new ideas.' As John Wheeler once said: 'Behind it all is surely an idea so simple, so beautiful, that when we grasp it – in a decade, a century, or a millennium – we will all say to each other, how could it have been otherwise?'

Berman points out that while the anomalous motion of Uranus was explained by Le Verrier's prediction of Neptune, the anomalous motion of Mercury was not explained by his prediction of Vulcan. It required a new idea: a fundamental change to gravity. 'Dark matter might be there and responsible for the anomalous motions of stars and galaxies,' says Berman. 'Or we've got to change gravity.'[37]

Is a big idea missing?

At this very moment there could be another Einstein out there who is in possession of the missing idea which will pull everything together and single-handedly create a new revolution in physics. But history suggests that a lone genius may be insufficient.

Einstein's theory of relativity was certainly the product of a lone genius – although Einstein himself remarked: 'I'm no Einstein.' But Arkani-Hamed points out that other revolutions in physics have taken more than one person. Quantum theory, for instance, was the work of about twenty physicists over a span of about twenty-five years. The Standard Model of particle physics required a similar number of people over a similar time span. The odds are therefore that the deeper theory than general relativity will be more like these revolutions than the Einsteinian revolution, and future historians of science will not talk of Newton, Einstein and a third name.

Arkani-Hamed is expecting a revolution in our picture of the world more profound than the quantum revolution of the 1920s. In fact, he draws a parallel with the birth, development and crystallisation of quantum theory. The first hint of the new worldview came with Planck's discovery of the quantum in 1900. Later, in 1913, the Danish physicist Niels Bohr used the quantum to explain the atom in an ad hoc manner. Finally, by 1927, came the creation of a self-consistent quantum theory built on solid foundational principles. 'At this moment I think we are about halfway to our ultimate destination,' says Arkani-Hamed. 'In quantum terms, we have reached around 1917 to 1918.'

The undiscovered country

'This is the most exciting time to be doing physics since the 1920s,' says Arkani-Hamed. 'Every generation since the ancient Greeks could have asked: "Where did the Universe come from?" and "What are space and time?" But all previous generations had a lot of other questions to answer before they could get to address these big ones. We've answered them. Now the big questions are the next questions.'

According to Arkani-Hamed, this is a singular moment in the history of fundamental physics. For the first time in history, we have a framework which allows us to ask the big questions and fantastic experimental probes such as the LHC to help us answer them. 'We've got to base camp at Everest,' says Arkani-Hamed. 'Before us we can see the beast.'

How long until we get to our destination? 'Maybe we need the results from five experiments, that's all,' says Arkani-Hamed. 'On the other hand, it could take us 500 years. But I don't think so. I'm more optimistic than that.'

The deeper theory will tell us about the birth of the Universe. It will tell us about where space and time and *everything* came from, and most importantly *why* they exist. But it will also tell us, in the words of Einstein, 'whether God had any choice in the creation of the World'.

But such a theory, as well as telling us profound things about the world we live in, may also give us technological mastery over that world. Maxwell's unification of electricity and magnetism in 1863 led ultimately to both special relativity and quantum theory. Quantum theory can be said to have created the modern world, giving us lasers and computers, iPhones and nuclear reactors. Inventions which exploit quantum theory are estimated to account for about 30 per cent of the GDP of the United States.

Maxwell's theory also predicted the existence of radio waves and led directly to our connected world, in which data and moving pictures and the invisible chatter of billions continually course through the air all around us. Neither Maxwell nor any of his contemporaries predicted any of this. If people in the nineteenth century could have seen television or the Internet or mobile phones, they would probably have considered them less technological artefacts than devilish manifestations of the supernatural.

Who knows what the deeper theory than Einstein's will give us? 'I defy gravity,' said Marilyn Monroe. And maybe we will too. Perhaps we will gain mastery over space and time, the ability to create wormholes, to build star ships or to fabricate time machines. 'We might learn how to create universes in the lab,' says Arkani-Hamed.

'Nothing', as Michael Faraday remarked, 'is too wonderful to be true.'

'Whether you can go back in time is held in the grip of the law of quantum gravity,' says Kip Thorne. 'We are several decades away from a definitive understanding, 20 or 30 years, but it could be sooner than that.'[38]

'The rapid progress true science now makes, occasions my regretting sometimes that I was born so soon,' wrote Benjamin Franklin. 'It is impossible to imagine the height to which may be carried in a thousand years the power of man over matter. We may perhaps learn to deprive large masses of their gravity and give them absolute levity, for the sake of easy transport.'[39]

As the example of Maxwell's theory shows, the spin-offs of the deeper theory are likely to be as stupendous as they are unguessable. Science-fiction writer Arthur C. Clarke put it best when he said: 'Any sufficiently advanced technology will be indistinguishable from magic.'[40]

Brace yourself for the magical world just over the horizon. Who knows what we will find in the undiscovered country?

Notes

Chapter 1

1   1714, 'Portsmouth Collection' of Newton's papers, 1714.
2   Elizabeth Knox, *The Vintner's Luck*, Vintage, London, 2000.
3   William Stukeley, *Memoirs of Sir Isaac Newton's Life*, 1752, pp. 46–9.
4   Fouad Ajami, 'The Arab World's Unknown Son', *Wall Street Journal*, 12 October 2011.
5   Daniel Defoe, *Journal of the Plague Year*, 1722.
6   'He died on March the 20th, 1727, after more than eighty-four years of more than average bodily health and vigour; it is a proper pendant to the story of the quart mug to state that he never lost more than one of his second teeth' – Augustus De Morgan, *Essays on the Life and Work of Newton*, 1914.
7   Stukeley, *Memoirs of Sir Isaac Newton's Life*, pp. 46–9.
8   Richard Westfall, *Never at Rest: A Biography of Isaac Newton*, 1983, p. 53.
9   Defoe, *Journal of the Plague Year*.
10 William Wordsworth, *The Prelude*, 1888.
11 The planets crawl across the night sky in a narrow band named the Zodiac along which there are twelve prominent groups of stars, or 'constellations', corresponding to the 'twelve signs of the Zodiac'. The reason for this is that the orbits of the planets are confined more or less to a single plane, known as the 'ecliptic'. And the reason for this is that the planets all formed from a flat disc of debris swirling around the newborn Sun.
12 The stars appear fixed relative to each other simply because of their immense distances from us. Travelling to even the nearest would be equivalent to flying a *billion* times around the Earth. But the stars *are* flying through space, and, over very long periods of, say, tens of thousands of years, they do move significantly enough to alter some of the constellations beyond recognition.
13 The Solar System is defined as the Sun plus its planets and moons plus the assorted debris – asteroids and comets – left over from its formation 4.55 billion years ago.
14 W. W. Rouse Ball, *History of Mathematics*, 1901.

15   This aspect of Newton's character was noted by a twentieth-century
       biographer, John Maynard Keynes: 'His peculiar gift was the power of
       holding continuously in his mind a purely mental problem until he had
       seen it through.' Keynes, 'Newton, the Man'. In *Essays in Biography*,
       1933.

16   The area of a small triangle swept out by a planet in a given time
       is ½vrt. The fact that ½vrt does not change is telling us that mvr,
       the planet's 'angular momentum', does not change either. This can
       happen only if there is no turning force, or 'torque' – no force along
       the trajectory of the planet. In other words, the force must always be
       directed towards the Sun.

17   Even today it is a mystery why the Universe appears to be describable
       by mathematical formulae. The twentieth-century Austrian physicist
       Eugene Wigner remarked on 'the unreasonable effectiveness of
       mathematics in the physical sciences'. Why is there a mathematical
       world which is a perfect analogue of the real world? No one knows.

18   Laws of Physics for Cats (http://www.funny2.com/catlaws.htm).

19   The planets are moving because they were set in motion around the
       Sun when the Solar System was born and, ever since, have just *kept*
       on going. In the modern picture the Sun and planets formed from an
       interstellar cloud of dust and gas, which began shrinking under its own
       gravity. The cloud would have been rotating a little because our Galaxy,
       the Milky Way, is rotating – once every 220 million years – and that
       rotation would have been amplified as the cloud shrank, just as a ballet
       dancer's spin quickens as she pulls in her arms. Inevitably, the planets
       that formed within the cloud, from the accretion of debris, would have
       inherited the rotational motion, and thus been born moving around the
       new-born Sun.

20   Actually, a little bit of simple reasoning can yield the *exact* form of
       the centripetal force. If a body is moving slowly in a circle, it needs
       only a small velocity correction towards the centre to stop it flying off
       on a tangent; if it is moving fast, it requires a big velocity correction.
       So the velocity correction goes up with the body's velocity (it is
       proportional to v). Now, the 'acceleration' of a body is how fast its
       velocity is changing, which is its velocity change in a given time. The
       time the body takes to cover a given distance is obviously shorter if
       the circle is small and longer if it is moving slowly (it is proportional
       to r/v). Consequently, the acceleration is proportional to v divided by
       r/v, which is v^2/r. And the force, which is simply the mass times the
       acceleration, is mv^2/r.

21    $mv^2/r = F(r)$. $T^2 \sim r^3 => v^2 \sim 1/r$. And so $F(r) \sim 1/r^2$. (Here m stands for
       the mass of a planet; v for its velocity, F the force of gravity exerted on
       it by the Sun, and r the distance of the planet from the Sun.)

22   In fact, there is one significant anomaly concerning the orbits of

Jupiter's moons. It was discovered by Ole Christensen Rømer in 1676. The Danish astronomer had spied the moons go around Jupiter many times and timed how long on average it takes each to make a complete orbit – since they periodically go behind Jupiter, the moment they re-emerge is a good moment to begin timing. Röemer was surprised to discover that the moons sometimes appear to be *ahead* of schedule and sometimes *behind*. They are ahead when Jupiter is at its closest to the Earth and behind when Jupiter is at its furthest away. What is going on? Rømer realised that it takes time for the light from Jupiter's moons to cross the space from Jupiter to the Earth. And when Jupiter is at its furthest from us the light takes longer than when it is at its nearest. This is why the moons emerge from behind Jupiter early or late, depending on whether they are closer to or further from the Earth. The phenomenon shows that light does not travel instantaneously. Furthermore, by knowing the extra distance the light has to travel when Jupiter is far away – the diameter of the Earth's orbit – and the time delay – 22 minutes, Rømer was able to make the first-ever estimate of the speed of light. He got 225,000 kilometres per second. Not bad considering that the modern estimate is 299,792 kilometres a second. The reason for Rømer's error was that his estimate of the size of the Earth's orbit was wrong and the time delay is not 22 minutes but 16 minutes and 40 seconds.

23 Douglas Adams, *Life, the Universe and Everything*, Picador, London, 2002.

24 The first person to determine accurately both the size and distance of the Moon was the Greek astronomer, geographer, and mathematician Hipparchus, who lived between 190 and 120 BC. During a lunar eclipse he estimated the size of the Earth's shadow on the Moon, and found it to be 2.5 times the diameter of the Moon. He realised, correctly, that, because the shadow is cast on the *curved* surface of the Moon, it was shrunk – by 1 Moon diameter – so the Earth is actually 3.5 times the diameter of the Moon. So, if the Earth were viewed from the same distance as the Moon it would appear 3.5 times bigger – that is, instead of appearing only about 0.5 degrees across, as the Moon does, it would appear 1.75 degrees across. (Your thumb held at arm's length will just cover the Moon – that is, it 'subtends' about 0.5 degrees.) The only way something that is the diameter of the Earth can appear 1.75 degrees across in the sky is if it is about 30 Earth diameters away. So the distance between the Earth and Moon is 384,400 kilometres. Obviously, Hipparchus did not get precisely this figure. But he was close.

25 The diameter of the Earth was first estimated by Eratosthenes, chief librarian of the Museum at Alexandria, in 240 BC. Actually, apart from wrinkles like mountains, the Earth seems flat. But, as

Eratosthenes realised, this is because the Earth is big and its curvature imperceptible. Evidence that the Earth is round comes from ships at sea, which disappear over the horizon while still sizeable whereas they should dwindle to dots first if the Earth were flat. Also, during a lunar eclipse, when the Earth passes between the Sun and Moon, the Earth's shadow on the Moon is curved, and the only body that produces a curved shadow from every direction is a sphere. The cleverness of Eratosthenes was to notice that, at the summer solstice when the Sun reached its highest point in the sky, a vertical pillar at Syene (modern-day Aswan) had no shadow – the Sun was directly overhead. On the same day, a pillar at Alexandria had a short shadow, revealing that the Sun was 7 degrees from the vertical. Knowing the separation of Syene and Alexandria, and that 7 degrees is about 1/50th of a full circle, Eratosthenes calculated the Earth's circumference. From this he obtained a diameter of 7,800 miles, which, incredibly, was only 100 miles out!

26 *It's Only A Theory*, BBC4, 2009.

27 A. C. Grayling, *The Good Book,* Bloomsbury, London, 2013.

28 *New York Post*, 24 June 1965.

29 'Fragments from a Treatise on Revelation'. In Frank E. Manuel, *The Religion of Isaac Newton*, Oxford University Press, Oxford, 1974.

30 I speculate on why the Universe is simple in Chapter 6 of my book *The Never-Ending Days of Being Dead*, Faber & Faber, 2007. In Chapter 2 of the same book, I speculate on why the simplicity may be an illusion, caused by physicists focusing only on the simple bits of the Universe! And in Chapter 8 of my book *The Universe Next Door*, Headline, 2002, I speculate on why the Universe is mathematical.

31 Isaac Newton, Query 31, *Opticks*, 1730.

32 'The Potentialities and Limitations of Computers', lectures by Richard Feynman and Gerry Sussman attended by the author, California Institute of Technology, Pasadena, 1984.

33 Actually, in Germany, the mathematician Gottfried Leibniz had independently invented calculus. In fact, he had invented it before Newton published anything about it, though Newton claimed he knew it even earlier, in 1666, and had described it to Leibniz in correspondence. In his later incarnation as President of the Royal Society, Newton did everything in his power to crush his rival and take sole credit for the invention of calculus.

34 Peter Ackroyd, *Newton*, Vintage, London, 2007, p. 10.

35 George Gamow, *The Great Physicists from Galileo to Einstein*, Dover, New York, 1988.

36 Eyesight deteriorates with age for a variety of reasons and it is very likely that Newton's did too. Often, the deterioration takes the form of a cataract – a clouding of the eye's natural lens, which lies behind the

iris and the pupil. One type, known as a nuclear cataract, forms deep in the central zone, or nucleus, of the lens. When a nuclear cataract first develops, it can bring about a temporary improvement in near vision, dubbed 'second sight'. Such improved vision is short-lived and will disappear as the cataract worsens. So is the explanation of Newton's exceptional eyesight at eighty-four that he was benefiting from the temporary improvement of a nuclear cataract and fortunately died before the improvement disappeared?

37 Keynes, 'Newton, the Man'.

Chapter 2

1 Quoted in Forest Ray Moulton, *Introduction to Astronomy*, Macmillan, New York, 1906, p. 199.

2 In 1930, in an after-dinner toast to Albert Einstein, who was present. Quoted in Blanche Patch, *Thirty Years with G.B.S.*, Gollancz, London, 1951.

3 Hazel Muir, 'Einstein and Newton showed signs of autism', *New Scientist*, 30 April 2003 (https://www.newscientist.com/article/dn3676-einstein-and-newton-showed-signs-of-autism/).

4 Even Newton's 'reflecting telescope' was a spin-off of his monumental work on light and optics, which was also secret and was to be dragged out of him – pretty much everything had to be dragged out of him – and published as *Opticks* as late as 1710.

5 Abraham DeMoivre, quoted in Richard Westfall, *Never at Rest: A Biography of Isaac Newton*, Cambridge University Press, Cambridge, 1983.

6 More than three centuries later, another genius, the American physicist Richard Feynman, so wanted to get into the mind of Newton that he re-invented a geometrical proof that bodies under an inverse-square law force travel in ellipses. After Feynman's death in 1988, his friends David and Judith Goodstein published it in *Feynman's Lost Lecture: The Motion of the Planets Around the Sun*, Jonathan Cape, London, 1996.

7 Ibid.

8 Isaac Newton, *Philosophiæ Naturalis Principia Mathematica* (1687), 'General Scholium'.

9 Abdus Salam, C. H. Lai and Azim Kidwai, *Ideals and Realities: Selected Essays of Abdus Salam*, World Scientific, Singapore, 1987.

10 Sir David Brewster, *Memoirs of the Life, Writings, and Discoveries of Sir Isaac Newton*, 1855.

11 Peter Ackroyd, *Newton*, Vintage, London, 2007, p. 29.

12 James Gleick, *Isaac Newton*, HarperCollins, London, 2004, p. 8.

THE ASCENT OF GRAVITY

Chapter 3

1   William Shakespeare, *Julius Caesar*, Act IV, Scene 3.

2   Although this saying is often attributed to Geoffrey Chaucer, its
      first apppearance in this form dates from the eighteenth century. It
      is listed as an existing proverb by Nathan Bailey in his *Dictionarium
      Britannicum: Or, A More Compleat Universal Etymological English
      Dictionary Than any Extant* (Second edition, 1736).

3   The word *bore* derives from the Old Norse word *bára*, meaning 'wave'
      or 'swell'.

4   The bore can be as much as 7.5 metres high and reach a speed of 27
      kilometres an hour.

5   Kieran Westley and Justin Dix, 'Coastal environments and their role in
      prehistoric migrations', *Journal of Marine Archaeology*, vol. 1, 1 July
      2006, p. 9 (http://www.science.ulster.ac.uk/cma/slan/westley_dix_2006.
      pdf).

6   Julius Caesar, 'Caesar in Britain. Heavy Damage to the Fleet', *History
      of the Gallic Wars*.

7   Martin Ekman, 'A concise history of the theories of tides, precession-
      nutation and polar motion (from antiquity to 1950)', 1993 (http://
      www.afhalifax.ca/magazine/wp-content/sciences/vignettes/supernova/
      nature/marees/histoiremarees.pdf).

8   The tidal force on the ocean furthest from the Moon is slightly smaller
      than on the ocean nearest the Moon by a factor of about $(60/62)^2 =$
      0.94 because the ocean there, rather than being 60 Earth radii from the
      Moon is 62 Earth radii from the Moon. Consequently, the tidal bulge is
      slightly smaller.

9   From the observation that the tides pulled by the Moon are about
      twice as big as those pulled by the Sun, Newton was able to deduce
      that the average density of the Moon is about twice that of the Sun.
      His logic is as follows: the tidal force exerted by a body depends on its
      mass. Tidal forces are also due to differences in gravity so they weaken
      not according to an inverse-square law but an inverse-cube law. The
      tidal force exerted by a mass m at a distance r is, therefore, $\sim m/r^3$. But
      $m \sim \rho d^3$, where ρ is its average density and d is its diameter. d is just
      $r\theta$, where θ is the angle subtended by the body in the sky. Putting all
      this together implies that the tidal force $\sim \rho \theta^3$. But the angular size of
      the Moon and Sun, by a cosmic coincidence, are almost the same – it
      is why the Moon can exactly blot out the Sun during a total eclipse.
      Consequently, the Moon and Sun exert a tidal effect *in proportion to
      their densities* – a surprising result. Since the Moon pulls tides twice as
      big as the Sun, it follows that it must have twice the average density of
      the Sun.

10   The plane of the Moon's orbit is inclined to the Earth's equator,
      varying between 18.28 and 28.58 degrees of the equatorial plane.

11   To be precise, maximum bores occur one to three days after new and full moons.

12   Chaim Leib Pekeris, 'Note on Tides in Wells', *Travaux de l'Association Internationale de Géodésie*, Paris, vol. 16, 1940.

13   In 1905, Albert Einstein discovered that mass is simply a super-compact form of energy (his formula $E = mc^2$ describes the exact connection, with c representing the speed of light). According to the law of conservation of energy, energy cannot be created or destroyed, only converted from one form to another. This means that the energy of motion (kinetic energy) of colliding subatomic particles can be converted into the mass-energy of new particles. This, in a nutshell, is how particle colliders such as the one at CERN work.

14   Technically, the protons have an energy of 7 teraelectronvolts (TeV), giving a total collision energy of 14 TeV. At 99.9999991 per cent of the speed of light, they travel around the CERN ring 11,000 times a second. Their 'Lorentz factor', γ, is about 7,500, which means they are 7,500 times more massive than protons at rest. This is an effect of Einstein's special theory of relativity, which ensures massive bodies become ever more massive and ever harder to push as they approach the speed of light, so that the speed of light is forever unreachable (see Chapter 5). Although the LHC protons travel within a mere 3 metres per second of the speed of light – about the speed of a jogger – boosting their velocity by that amount would require an infinite amount of energy.

15   Every subatomic particle has an antimatter twin with opposite properties such as electric charge and quantum 'spin'. The antiparticle of the negatively charged electron is the positively charged positron.

16   L. Arnaudon et al., 'Effects of terrestrial tides on the LEP beam energy', *CERN SL/94-07 (BI)*, 1995 (https://jwenning.web.cern.ch/jwenning/documents/EnergyCal/tide_slrep.pdf).

17   To keep a body of mass, m, moving in a circle of radius, r, at a velocity, v, requires a centrally directed 'centripetal force', $F = mv^2/r$ (see Chapter 1). If the radius of the ring gets bigger, the force, F, exerted by the LEP magnets, which is constant, is too big to keep the particles travelling around the larger circle, unless v^2, which is related to the energy of the particles, goes up in the same proportion. On the other hand, if the radius of the ring gets smaller, the force, F, exerted by the magnets is too small to keep the particles travelling around the smaller circle, unless the energy of the particles goes down in the same proportion.

18   The tidal effect on CERN's accelerator ring is not the only effect that has been observed by the laboratory's physicists. Every day, at very particular times, the energy of the beams has to be corrected. It took the physicists many months to discover why. Bizarrely, it was the fast

TGV train linking Geneva and Paris. As it passed close to the LEP ring, it released a lot of electrical energy into the ground which perturbed the particle beams.

19   The same logic predicted that the tidal forces on the Mediterranean are less than half those on the Atlantic because the Mediterranean's depth is on average less than half that of the Atlantic.

20   Arlin Crotts, 'Transient Lunar Phenomena: Regularity and Reality', 2007 (http://xxx.lanl.gov/PS_cache/arxiv/pdf/0706/0706.3947v1.pdf).
     Arlin Crotts, 'Lunar Outgassing, Transient Phenomena and the Return to the Moon, I: Existing Data', 2007 (http://xxx.lanl.gov/PS_cache/arxiv/pdf/0706/0706.3949v1.pdf).
     Arlin Crotts and Cameron Hummels, 'Lunar Outgassing, Transient Phenomena and the Return to the Moon, II: Predictions of Interaction between Outgassing and Regolith', 2007 (http://xxx.lanl.gov/PS_cache/arxiv/pdf/0706/0706.3952v1.pdf).
     Arlin Crotts,'Lunar Outgassing, Transient Phenomena and the Return to the Moon, III: Observational and Experimental Techniques', 2007 (http://xxx.lanl.gov/PS_cache/arxiv/pdf/0706/0706.3954v1.pdf).
     Marcus Chown, 'Does the Moon have a volcanic surprise in store?' *New Scientist*, 26 March 2008.

21   The creation of the Mare basins is associated with the Late Heavy Bombardment. This is believed to have happened when Jupiter and Saturn, in moving to their present locations, briefly entered a 2:1 resonance in which for every two orbits Jupiter made around the Sun, Saturn circled just once. This periodically brought the two planets close together, boosting their gravitational effect on other bodies. Like a child periodically pushed on a swing that gets ever higher, small bodies such as the rocky asteroids were pushed ever more from their orbits, and plunged into the inner Solar System, where they bombarded the inner planets such as the Earth and Moon.

22   L. Chen et al., 'Correlations between solid tides and worldwide earthquakes MS ≥ 7.0 since 1900', *Natural Hazards & Earth System Science*, vol. 12, 2012, p. 587.

23   Actually, because of a wobble caused in the motion of the Moon known as 'libration' and the fact we see the Moon from different directions depending where we are on the planet – an effect known as 'parallax' – we see about 59 per cent of the lunar surface.

24   The tidal bulge makes an angle of 3 degrees with the Moon, so there is a 3/360 hours × 24 hours = 12 minutes difference between the time a high tide is expected to arrive and the time it actually arrives.

25   Adam Hadhazy, 'Fact or Fiction: The Days (and Nights) Are Getting Longer', *Scientific American*, 14 June 2010.

26   Marcus Chown, 'In the shadow of the Moon', *New Scientist*, 30 January 1999.

27 The angular momentum of a point mass, m, is defined as its linear momentum, mv, multiplied by the distance, r, from the centre of rotation. Since the orbital velocity of a body at a distance, r, from the Earth is proportional to $1/r^{\frac{1}{2}}$, this means the angular momentum is proportional to $r \times 1/r^{\frac{1}{2}} = r^{\frac{1}{2}}$. So the angular momentum of the Moon does indeed go up as it recedes from the Earth.

28 The Lunokhod 2 reflector works occasionally but the one on Lunokhod 1 was lost for almost forty years. But recently the Lunar Reconnaissance Observer probe imaged the landing site. The coordinates were passed to scientists in New Mexico. And, remarkably, they fired a pulse of laser light at the landing site and, on 22 April 2010, were stunned to receive a return burst of 2,000 particles of light, or 'photons'. With four, and possibly five, corner-cubes now in action it will be possible to observe not only the recession of the Moon but changes in its shape as it is tidally stretched and squeezed by the Earth.

29 J. O. Dickey et al., 'Lunar Laser Ranging: A Continuing Legacy of the Apollo Program', *Science*, vol. 265, 1994, p. 482.

30 In a system of two large bodies bound together by gravity, the Lagrange points are locations at which the combined gravitational pull of the two bodies provides precisely the centripetal force (see Chapter 1) required to orbit with them. There are five such points, which are labeled L1 to L5.

31 J. Green and Matthew Huber, 'Tidal dissipation in the early Eocene and implications for ocean mixing', *Geophysical Research Letters*, vol. 40, 2013.

32 The Sun is actually using just about the most inefficient nuclear reaction imaginable. It is turning 'nuclei' of the lightest element, hydrogen, into nuclei of the next heaviest, helium. Hydrogen consists of 1 nuclear Lego brick and helium 4, so 'hydrogen-burning' is a multi-step process. The first step is the 'fusion' of two hydrogen nuclei, or protons. But, on average, it takes two protons in the Sun 10 billion years to meet each other and stick. This is the reason the Sun will take about 10 billion years to burn its hydrogen fuel – it is about half way through its life – and there has been sufficient time for the evolution of complex life like us. The Sun is so inefficient at generating heat that, if you were to take your stomach and a piece of the core of the Sun the same size and shape as your stomach, your stomach would generate more heat. You might then ask: how come the Sun is so hot? The answer is that the Sun does not simply contain one chunk of matter the size and shape of your stomach; it contains countless quadrillion chunks, all stacked together.

Chapter 4

1 Isaac Newton, *The Principia*, edited by Florian (1687).

2   Despite research I have been unable to find the origin of this quote,
     which is widely attributed to Paul Dirac.

3   A modern example of the scarcely believable, predictive power of
     science is the discovery of the 'Higgs particle' in July 2012. Hiking in
     the Cairngorm Mountains of Scotland in 1964, Peter Higgs realised
     that the fundamental building blocks of all matter must gain their
     mass by interacting with a kind of invisible treacle, now called the
     'Higgs field', that fills all of space, and that, furthermore, a localised
     excitation of that field should manifest itself as a new subatomic
     particle. (To be fair, Higgs was one of five physicists to independently
     come up with the 'Higgs mechanism'. But his is the name that has
     stuck.) Almost four decades later, at a cost of more than 10 billion
     euros, the biggest machine in the world – the Large Hadron Collider
     near Geneva – found the 'Higgs particle'. It is just as profound a shock
     to physicists today, as it was to those in Le Verrier's time, that nature
     dances to the tune of the arcane mathematical equations they scrawl
     across pieces of paper.

4   In physics, the only system that is 'exactly soluble' – that is, whose
     evolution can be deduced for all time – is the two-body system.
     This includes the Earth and Moon moving under the influence of
     their mutual gravitational force and the proton and an electron of a
     hydrogen atom moving under the influence of their mutual electric
     force. Once a third body is introduced, things become so horrendously
     complicated that approximation is the best mathematicians can do. (To
     calculate the trajectory of an interplanetary space probe, for instance,
     mission planners have to resort to a 'brute force' method. At each
     location, they have to sum up the forces on the space probe from all the
     planets, determine how it moves in response to that sum over the next
     minute; then repeat the whole process at the new location where the
     forces from all the planets will be slightly different; and so on.) In fact,
     the long-term evolution of a system of three or more masses under
     the influence of their mutual gravity, though predictable in theory,
     is unpredictable in practice. Because of this phenomenon, known as
     'deterministic chaos', even a tiny difference in the starting locations
     of planets will, after a while, result in wildly differing behaviour in
     the distant future. Even worse, the Solar System is unstable in the long
     term. Like a clock mechanism that unpredictably goes wild, its cogs
     and wheels flying off in all directions, the Solar System could one day
     eject Mercury or Mars or any other body. In fact, in the distant past, it
     may have flung a planet or two into the frigid darkness of interstellar
     space.

5   Caroline Herschel has the distinction of discovering more comets than
     any other woman apart from another Caroline, Caroline Shoemaker,
     in the late twentieth and early twenty-first centuries.

6   William Sheehan and Steven Thurber, 'John Couch Adams's Asperger syndrome and the British non-discovery of Neptune', *Notes and Records of the Royal Society Journal of the History of Science*, vol. 61, issue 3, 22 September 2007 (http://rsnr.royalsocietypublishing.org/content/61/3/285).

7   The planets all orbit in the same plane as if they are confined to a giant transparent dinner plate centred on the Sun. This is because of the way in which the Solar System formed 4.55 billion years ago. A spherical cloud of gas and dust shrank under its own self-gravity. Because it was spinning – our Milky Way is a spinning whirlpool of stars, so it is plausible that it was – the cloud shrank faster between its poles than it did around its waist, where gravity was counteracted by a tendency of the cloud material to be flung outwards. As a result, the spherical cloud collapsed to form a thin disc of gas and dust swirling around the newborn Sun. It is because the planets were built up from rubble that collided and stuck together within this disc that they orbit in roughly the same plane and also go around the Sun in the same direction.

8   See Chapter 3.

9   See Chapter 3.

10   Konstantin Batygin and Mike Brown, 'Evidence for a distant giant planet in the Solar System', *Astronomical Journal*, vol. 151, 20 January 2016, p. 22.

11   A planet shines principally by reflected light from its star, whereas a star generates its own light via nuclear reactions in its core. Such nuclear reactions require a temperature of millions of degrees to ignite, which in turn requires a lot of mass to be bearing down on the star's core – when things are squeezed they get hot, as anyone who has squeezed the air in a bicycle pump knows. The threshold that divides planets from stars is about 0.08 times the mass of the Sun, or about 80 times the mass of Jupiter. Bodies less massive than this are planets, objects more massive are stars.

12   A spectrograph uses a 'diffraction grating' to fan starlight out into its constituent rainbow colours. A grating often consists of many parallel scratches on the surface of a flat piece of transparent material, and is superior to a glass 'prism'. Atoms of a particular element in the outer atmospheres of stars create dark bands at characteristic frequencies. Measuring the Doppler shift simply involves seeing how far such bands are shifted in frequency from similar bands created by their earthbound cousins.

13   Exoplanets are not only found by observing a 'wobble' in their parent star. If a planet orbits a star in such a way that it passes periodically across the face of the star as seen from the Earth, then it can dim the light of the star, by about 1 per cent for a Jupiter-mass planet and 0.01 per cent for an Earth-mass planet. The Kepler space observatory,

launched into Earth orbit in 2009, has monitored the light of more
than 100,000 stars, and found more than 1,000 exoplanets this way.

14   Not everyone has an unshakeable faith in Newton's law of gravity.
     A sizeable minority of astronomers, led by Mordehai Milgrom of
     the Weizmann Institute in Rehovot, Israel, believe that, below an
     acceleration of about one-billionth of a g, gravity changes to a stronger
     form that does not weaken as quickly with distance as an inverse-
     square-law force. This Modified Newtonian Dynamics, or MOND, can
     describe the motions of stars orbiting in all spiral galaxies with a single
     formula. By comparison, a different amount of dark matter with a
     different distribution is required to explain the motion of stars in each
     spiral galaxy. A form of MOND which is compatible with Einstein's
     theory of relativity was found by Jacob Bekenstein of the Hebrew
     University of Jerusalem in 2000. See Jacob Bekenstein, 'Relativistic
     gravitation theory for the MOND paradigm' (http://arxiv.org/pdf/
     astro-ph/0403694v6.pdf).

15   Vera Rubin, N. Thonnard and Kent Ford, 'Rotational Properties of 21
     Sc Galaxies with a Large Range of Luminosities and Radii from NGC
     4605 (R=4kpc) to UGC 2885 (R=122kpc)', *Astrophysical Journal*, vol.
     238, 1980, p. 471 (http://adsabs.harvard.edu/abs/1980ApJ...238..471R).

16   For more about the Large Synoptic Survey Telecope, see http://www.
     lsst.org/

17   See Marcus Chown, *Afterglow of Creation*, Faber & Faber, London,
     2010.

18   Ibid.

19   A black hole is a region of space where gravity is so strong that nothing
     can escape, not even light, hence its blackness. We have discovered two
     types. There are stellar-mass black holes, formed when gravity crushes
     a massive star out of existence at the end of its life. And there are
     'supermassive' black holes, up to 50 billion times as massive as the Sun,
     of unknown origin, which lurk in the heart of every galaxy, including
     our own. But some physicists suggest that there could be a third type
     of black hole: miniature versions created in the first split-second of the
     big bang and which have survived until the present day.

20   Table 2 here shows the perihelion precession rates of the eight planets
     of the Solar System (http://farside.ph.utexas.edu/teaching/336k/
     Newtonhtml/node115.html).

21   Ceres, the largest asteroid, visited by NASA's 'Dawn' spacecraft in
     2015, was discovered on the first day of the nineteenth century. It was
     followed in 1807 by Vesta, and then a host of others. Initially, Ceres
     was thought to be a new planet. But the combined mass of all the
     hundreds of thousands of asteroids is barely 1 per cent of the mass
     of the Earth. The asteroids are believed to be builder's rubble left over
     from the birth of the Solar System. They were unable to aggregate

together to form a planet because of the disruptive gravitational effect of nearby Jupiter. Ceres is now classified as one of the Solar System's five 'dwarf planets'.

22   Sunspots are regions of the Sun where intense loops of magnetic field burst through the 'surface', or photosphere. The outward pressure of hot gas within a sunspot need not be as great as elsewhere since it is supplemented by the outward pressure of the magnetic field. Consequently, the gas is a couple of thousand degrees cooler than the 5,800 degrees Celsius temperature of its surroundings. It is because sunspots are cooler than average that they appear black. See Lucie Green, *15 Million Degrees*, Viking Penguin, London, 2016.

Chapter 5

1   Roberto Trotta, *The Edge of the Sky*, Basic Books, New York, 2014.

2   Albert Einstein, 'On the electrodynamics of moving bodies', *Annalen der Physik*, vol. 17, 1905, pp. 891–921. Completed June 1905, received 30 June 1905.

3   'During this year in Aarau, the following question occurred to me: If one pursues a beam of light with the velocity c (velocity of light in a vacuum) one should observe such a beam of light as a spatially oscillatory electromagnetic field at rest. Unfortunately, there seems to be no such thing! This was the first childlike thought experiment that was concerned with the special theory of relativity . . .' Albert Einstein, *Autobiographische Skizze*. In Carl Seelig (ed.), *Bright Times – Dark Times*, Europa Verlag, Zurich, 1956, p. 146.

4   Thomas Young, who lived in London, may have noticed raindrops falling on a puddle and the way in which concentric ripples spread out from the impact sites and overlap, reinforcing wherever two crests coincide and cancelling out wherever a crest and a trough coincide. If a vertical barrier were placed across the puddle, locations on the barrier hit by strong ripples would alternate with places where the water is calm. Young reasoned that, if he could demonstrate a similar 'interference' effect with light, he would prove that it is a wave. He shone a light on a screen with two vertical slits. On the far side, there emerged concentric ripples of light. In the region where they overlapped, he placed a vertical white barrier. Instantly, he saw a pattern of illuminated and dark bands, reminiscent of a supermarket barcode. He had proved light was a wave. Not only that but the spacing of the bands enabled him to deduce that its 'wavelength' – the distance over which it makes a complete up-and-down oscillation – is less than a thousandth of a millimetre.

5   Charles Darwin, *On the Origin of Species*, 1859.

6   Actually, a connection between electricity and magnetism and light had been suspected earlier by Michael Faraday. In a letter dated

13 November 1845, he wrote: 'I happen to have discovered a direct relation between magnetism and light, also electricity and light, and the field it opens is so large and I think rich' (*The Letters of Faraday and Schoenbein, 1836–1862*, 1899, p. 148). Among other things, Faraday had found that a magnetic field could change the vibration plane, or 'polarisation', of a light wave, a phenomenon now known as 'Faraday rotation'.

7   Maxwell's equations predict the existence of a whole 'spectrum' of invisible-to-the-naked-eye electromagnetic waves, of which the waves of visible light are but a tiny portion. 'Radio waves' have a wavelength more than 1,000 times longer than visible light.

8   Richard Feynman, Robert Leighton and Matthew Sands, *The Feynman Lectures on Physics, Volume II*, Addison-Wesley, Boston, 1989, pp. 1–11.

9   The aether had to be stiff enough to ripple at the enormous frequency of a light wave yet insubstantial enough not to noticeably impede the orbit of planets around the Sun. This meant it had to be much stiffer than steel yet much thinner than air. No wonder physicists were in trouble trying to imagine it!

10   It was Einstein's friend, Marcel Grossman, a mathematics student he had met when a student in Zurich, who was instrumental in Einstein getting his job at the Patent Office. Grossman had spoken to his father, who had recommended Einstein to Friedrich Haller, the director of the Bern Patent Office. Even at the end of his life, Einstein wrote of his gratitude for what Grossman had done for him.

11   Abraham Pais, *Subtle is the Lord*, Oxford University Press, Oxford, 1982.

12   Albert Einstein, 'On a heuristic viewpoint concerning the generation and transformation of light', *Annalen der Physik*, vol. 17, 1905, pp. 132–84. Completed 17 March 1905, received 18 March 1905.

13   Albert Einstein, 'On a new determination of molecular dimensions'. Doctoral thesis. Completed 30 April 1905.

14   Albert Einstein, 'On the movement of particles suspended in fluids at rest, as postulated by the molecular theory of heat', *Annalen der Physik*, vol. 17, 1905, pp. 549–60. Completed May 1905, received 11 May 1905.

15   Albert Einstein, 'On the electrodynamics of moving bodies', *Annalen der Physik*, vol. 17, 1905, pp. 891–921. Completed June 1905, received 30 June 1905.

16   Albrecht Fölsing, *Albert Einstein*, Penguin Books, London, 1997, p. 53.

17   Kyoto lecture, 14 December 1922. See *Physics Today*, August 1982, p. 46.

18   Ibid.

19   Douglas Adams, *Mostly Harmless*, Pan, London, 2009.

20   Although relativity predicts that someone moving relative to you should appear to shrink in the direction of their motion, this is not what you would see because another effect is at play. Light from more distant parts of the person takes longer to reach you than from closer parts. This causes them to appear to rotate. So, if their face is pointing towards you, you will see some of the back of their head. This peculiar effect is known as 'relativistic aberration', or 'relativistic beaming'.

21   Igor Novikov, *The River of Time*, Canto, Cambridge, 2001.

22   The Dutch physicist Hendrik Lorentz and the Irish physicist George FitzGerald had realised that bodies must appear to shrink in their direction of motion – an effect now known as 'Lorentz-FitzGerald contraction'. But, unlike Einstein, they had not seen it as an inevitable consequence of the principle of relativity and the principle of the constancy of the speed of light.

23   Although Einstein's theory was initially known as the 'theory of relativity', once he generalised and extended the theory in 1915, it became known as the 'special theory of relativity' to distinguish it from the 'general theory of relativity'.

24   The fact there is no aether was revealed observationally by the American physicists Albert Michelson and Edward Morley. In 1888, they measured the speed of light when the Earth, in its orbit around the Sun, was flying in the same direction as their light beam; and, six months later, when the Earth was moving in the opposite direction. Just as a boat sailing into a wind has a different speed to one with the wind at its back, they expected the speed of their light to depend on how it met the aether wind. To their utter amazement, they measured the same speed of their light in both cases. The speed of light was constant. For his work, Michelson won the 1907 Nobel Prize for Physics.

25   Actually, if the speed of light is independent of the speed of its source, it follows from the principle of relativity that its speed is also independent of the speed of an observer.

26   The American physicist John Wheeler said: 'Time is what stops everything happening at once.'

27   If the train is travelling at a velocity, v, it is a matter of simple geometry to work out that a clock on the train runs slower than one not on the train by a factor of $1/\sqrt{(1 - v^2/c^2)}$. It is also possible to deduce that a ruler on the train shrinks by a factor of $1/\sqrt{(1 - v^2/c^2)}$. The quantity $1/\sqrt{(1 - v^2/c^2)}$ is known as the Lorentz factor and is usually represented by the Greek symbol γ.

28   What does it mean to say that something happens at a particular time – say, someone strikes a match at 11 o'clock? It means, Einstein realised, that two events – the hands reaching the arrangement we refer to as 11 o'clock and the striking of the match – are simultaneous. But

consider someone striking the match at the centre of a train carriage which is travelling from left to right. Someone at the far-left end of the carriage will see the match ignite earlier than someone at the far-right because, in the time the light is travelling to them, the train will have moved forward, shortening the distance the light needs to travel. Since being able to agree on events being simultaneous is, according to Einstein, the very foundation of telling the time, not being able to do so means there can be no such thing as universal time everyone agrees on.

29 Max Flückiger, *Albert Einstein in Bern*, Verlag Paul Haupt, Bern, 1972, p. 158.

30 Quoted in Charles Misner, Kip Thorne and John Wheeler, *Gravitation*, W. H. Freeman, New York, 1973, p. 937.

31 'No time like the present': Marcus Chown, *The Never-Ending Days of Being Dead*, Faber & Faber, London, 2007.

32 Strictly speaking, momentum and energy.

33 Albert Einstein, 'Does the inertia of a body depend on its energy content?', *Annalen der Physik*, vol. 18, 1905, pp. 639–41. Received 27 September 1905.

34 The speed of light is uncatchable only for a body with mass. A massless particle – and a particle of light, or 'photon', is massless – can travel at the speed of light.

35 Einstein was not the one to coin the term 'relativity'. In fact, he did not like it. It was the great German physicist Max Planck, who at a meeting in Stuttgart on 19 September 1906, first spoke of 'relative theory'. This was gradually morphed by others into 'relativity theory'. It was not until 1911 that Einstein reluctantly used the term in the title of a paper, and even then in inverted commas. After a few more years, he accepted the inevitable and dropped the quotation marks.

36 Asked by William Gladstone, British Chancellor of the Exchequer, 'What is the practical use of electricity?' Faraday replied: 'Why, sir, there is every probability that you will soon be able to tax it.'

37 In general a 'field' is a physical quantity that has a value for each point in space and time. It could be something simple like a temperature, which merely has a magnitude, or it could be something more like a magnetic field, which has not only a magnitude but a direction in 3D space.

38 Conrad Habicht was a friend of Einstein's from his Zurich student days. The pair, together with Maurice Solovine, rather grandiosely called themselves the 'Olympia Academy'. They met in cafés and talked of the ideas they had absorbed from their readings of science, philosophy and literature.

Chapter 6

1   Quoted in Engelbert Schucking and Eugene Surowitz, *Einstein's Apple: Homogeneous Einstein Fields*, World Scientific, Singapore, 2015, p. 2.

2   Michio Kaku, 'A theory of everything?' (http://p-i-a.com/Magazine/Issue6/MichioKaku.htm).

3   *The Collected Papers of Albert Einstein, Volume 5: The Swiss Years, Correspondence, 1902–1914*, translated by Anna Beck (Princeton University Press, Princeton, 1995, p. 46). In the seven years from 1902 to 1909, Einstein assessed an estimated 2,000 patent applications, but his review of the AEG application is the only one that survives. Swiss bureaucracy ensured the destruction of all other examples of Einstein's expert opinions, despite his being a stellar figure in the world of physics after 1905.

4   'I was falling. Falling through time and space and stars and sky and everything in between. I fell for days and weeks and what felt like lifetimes across lifetimes. I fell until I forgot I was falling' – Jess Rothenberg, *The Catastrophic History of You and Me*, Penguin, London, 2012.

5   Fortunately, all lifts in operation are safety lifts. If the cable breaks, the lift jams in the shaft. Not very pleasant but rarely deadly for its occupants.

6   Ideally, this experiment should be done on an ice rink where friction with the ground does not complicate things!

7   1g = 9.8 metres per second per second. It is the acceleration that gravity causes at the Earth's surface. In other words, every second, a falling apple – or anything else – increases its speed by 9.8 metres per second.

8   For an acceleration as small as 1g, the effect would actually be so tiny as to be unnoticeable except with precision instruments.

9   You might think that this explanation of the slowing down of time in strong gravity is a trick because it uses a clock with a beam of light which travels horizontally between mirrors and not vertically between mirrors. But the reason for using a clock with a horizontal beam is so as to be able to define a tick at a constant height – that is, where gravity is constant.

10   James Chin-Wen Chou et al., 'Optical clocks and relativity', *Science*, vol. 329, 24 September 2010, p. 1,630.

11   David Berman, 'String theory: From Newton to Einstein and beyond' (https://plus.maths.org/content/string-theory-newton-einstein-and-beyond).

12   The ants-on-the-trampoline analogy of how gravity works is not perfect. A major flaw is that it implicitly uses gravity to explain gravity! After all, it is gravity that pulls the bowling ball downwards and creates the depression in the trampoline. Of course, the trampoline

could be floating in space in zero-gravity and the bowling ball could be electrically charged and pulled from one side of the trampoline by the electric force of another electrically charged body. But that would make the analogy complicated and confusing. Best to stick with the imperfect analogy and try to ignore the imperfection!

13   Kaku, 'A Theory of Everything?'

14   You may wonder why our natural motion is to head for the bottom of the valley of space-time centred on the Earth but that the Earth's natural motion is to circle the valley of space-time centred on the Sun. The reason is that the Earth is flying through space with an appreciable velocity and so cannot fall into the Sun, whereas we are not flying relative to the Earth.

15   Isaac Newton, *The Principia*, edited by Florian (1687), p. 643.

16   The first gravitational wave detector – a 2-metre-long, 1.4-tonne aluminium cylinder designed to ring like a bell when hit by a space-time ripple – was built by Joe Weber of the University of Maryland. His spurious claim, in the 1970s, that he had detected gravitational waves destroyed his scientific reputation but kick-started the whole field.

17   Dennis Overbye, 'Gravitational Waves Detected, Confirming Einstein's Theory', *New York Times*, 11 February 2016 (http://www.nytimes.com/2016/02/12/science/ligo-gravitational-waves-black-holes-einstein.html).

18   Janna Levin, *Black Hole Blues*, The Bodley Head, London, 2016.

19   Davide Castelvecchi, 'The black-hole collision that reshaped physics', *Nature*, 23 March 2016 (http://www.nature.com/polopoly_fs/1.19612!/menu/main/topColumns/topLeftColumn/pdf/531428a.pdf).

20   This is a personal recollection of mine. As a physics graduate student at the California Institute of Technology – co-constructor of LIGO along with the Massachusetts Institute of Technology – I remember attending a talk given by Drever at Caltech, probably in 1984.

21   Albert Einstein, *Autobiographische Skizze*. In Carl Seelig (ed.), *Bright Times – Dark Times*, Europa Verlag, Zurich, 1956, p. 11.

22   Alex Bellos, *Here's Looking at Euclid!*, Free Press, New York, 2010.

23   Einstein to Heinrich Zangger, 6 December 1917, *Collected Papers of Albert Einstein*, vol. 8, doc. 403, p. 411 (http://einsteinpapers.press.princeton.edu/).

24   Despite his key role in developing poison gases during the First World War, Fritz Haber won the 1919 Nobel Prize for Chemistry for the invention of the Haber-Bosch process for synthesising the ammonia used in fertilisers from hydrogen and atmospheric nitrogen.

25   Quoted in Lee Smolin, *Three Roads to Quantum Gravity*, Basic Books, London, 2000, p. 137.

26   Because energy curves space-time (that is, creates gravity) and curved

space-time contains energy, it follows that energy creates curvature, which creates more curvature, which creates yet more curvature, and so on. Consequently, the general theory of relativity reduces to Newton's theory of gravity when the energy contained in the curvature of space-time is small, so that the only significant term is the gravity created by mass-energy. And, of course, when all bodies are moving much slower than the speed of light.

27   Einstein letter to Conrad Habicht, Bern, 24 December 1907.

28   Abraham Pais, *Subtle is the Lord*, Oxford University Press, Oxford, 1983, p. 20.

29   Ibid., p. 257.

30   Einstein letter to Paul Ehrenfest, Berlin, 16 January 1916.

31   Simon Newcomb, *The Elements of the Four Inner Planets and the Fundamental Constants of Astronomy: Supplement to the American Ephemeris and Nautical Almanax for 1897*, Government Printing Office, Washington DC, 1895, p. 184.

32   In 1902, Simon Newcomb famously declared: 'Flight by machines heavier than air is unpractical and insignificant, if not utterly impossible.' The following year Orville Wright proved him wrong.

33   Dennis Overbye, 'A Century Ago, Einstein's Theory of Relativity Changed Everything', *New York Times*, 24 November 2015.

34   The 'stress energy tensor' is just a bag that holds a lot of information about what is present in space-time at some point: its energy density, its momentum densities, pressures, stresses and so on (http://pitt.edu/~jdnorton/teaching/HPS_0410/chapters/general_relativity/index.html).

35   Although Einstein was German-born and German-based, he was strictly speaking no longer a German citizen, having renounced his citizenship at the age of twenty-one in 1896.

36   A particle of light, known as a 'photon', has no intrinsic, or 'rest', mass (if it did, it could never travel at the speed of light). Its effective mass is entirely due to its energy or, more precisely, its 'energy-momentum'.

37   1 arc second is 1/60th of an arc minute, which is 1/60th of a degree. 1 arc second is consequently 1/3,600th of a degree.

38   Thomas Levenson, *The Hunt for Vulcan*, Head of Zeus, London, 2015, p. 161.

39   Ilse Rosenthal-Schneider, *Reality and Scientific Truth*, Wayne State University Press, Detroit, 1981, p. 74.

40   Charles Chaplin, *My Autobiography*, Penguin, London, 2003.

41   Simone Bertault, *Piaf*, Harper & Row, New York, 1972.

42   Pais, *Subtle is the Lord*, pp. 311–12.

Chapter 7

1   Gregory Benford, 'Leaping the Abyss', *Reason Magazine*, April 2002 (http://reason.com/archives/2002/04/01/leaping-the-abyss).

2   John Wheeler and Kenneth Ford, *Geons, Black Holes & Quantum Foam*, W. W. Norton, New York, 2000.

3   Karl Schwarzschild's solution was for a 'non-spinning' black hole. But all astronomical bodies are spinning. Nevertheless, it was not until 1963 – almost half a century after Einstein published his general theory of relativity – that the distortion of space-time by a realistic, 'spinning', black hole was deduced by the New Zealand physicist Roy Kerr.

4   Although John Wheeler is often credited with coining the term 'black hole', actually he merely popularised it. 'In the fall of 1967, [I was invited] to a conference . . . on pulsars,' he wrote. 'In my talk, I argued that we should consider the possibility that the center of a pulsar is a gravitationally completely collapsed object. I remarked that one couldn't keep saying "gravitationally completely collapsed object" over and over. One needed a shorter descriptive phrase. "How about black hole?" asked someone in the audience. I had been searching for the right term for months, mulling it over in bed, in the bathtub, in my car, whenever I had quiet moments. Suddenly this name seemed exactly right. When I gave a more formal Sigma Xi-Phi Beta Kappa lecture . . . on December 29, 1967, I used the term, and then included it in the written version of the lecture published in the spring of 1968.' Wheeler and Ford, *Geons, Black Holes & Quantum Foam*, p. 296.

5   Irrespective of the appearance of a star that shrank down to form a black hole, the resultant black hole is identical and characterised by just three things – its mass, how fast it is spinning and its electrical charge. Since big things tend to have an equal quantity of negative and positive charge, making them chargeless, in practice a black hole is characterised by only its mass and spin rate. The American physicist John Wheeler summarised this state of affairs as: 'A black hole has no hair.' In other words, there is nothing that can be learnt from observing the exterior of a black hole about the events that led to its birth.

6   Initially, Schwarzschild believed that the singularity was at the horizon of the black hole. But it turned out this was just an artefact of the coordinate system he used. The true singularity is at the heart of the hole.

7   See Chapter 8.

8   The Heisenberg Uncertainty Principle also makes atoms possible for, as Richard Feynman observed: 'Atoms are completely impossible from the classical point of view.' An electron in an atom orbits an 'atomic nucleus' like a planet around the Sun. According to the theory of electromagnetism, it should act like a tiny radio transmitter, radiating

away its orbital energy as electromagnetic waves and spiralling into the nucleus in less than a hundred-millionth of a second. It is prevented from doing so because the quantum wave of an electron cannot be squeezed into an arbitrarily small volume. Or, from the particle point of view, an electron squeezed close to the nucleus is like a bee in an ever-shrinking box, getting angrier and battering itself ever more violently against the walls of its prison.

9   Marcus Chown, *We Need to Talk About Kelvin*, Faber & Faber, London, 2010.

10   The Pauli Exclusion Principle makes possible the variety of atoms – nature's fundamental Lego bricks – and so is ultimately behind the complexity of the world. According to the theory of electromagnetism, all electrons in an atom, having radiated away their orbital energy, should crowd into the lowest-energy orbit, as close to the nucleus as possible. If this happened not only would the atoms of all ninety-two naturally occurring elements have the same size but they would all behave in the same way. This is because the behaviour of the atoms of an element is determined by the way in which the electrons are arranged. The Pauli Exclusion Principle dictates that they occupy 'shells' around the nucleus, with the exact number of electrons in the outer shell determining the way in which the atom links up with other atoms to form chemical compounds.

11   Electrons have intrinsic 'spin', something with no analogue in the everyday world. They are not actually spinning but behave as if they do. Imagine, however, that they are spinning. They are spinning at the lowest rate that nature permits, and there are two possible ways they can spin: clockwise or anticlockwise ('up' and 'down', in the physics jargon). This means that two electrons are not in the same state if they have opposite spins. The Pauli Exclusion Principle therefore permits two electrons in the same location – not one – to have the same velocity.

12   Why is it the electrons that hold up the star against gravity and not the atomic nuclei? The answer is that the nuclei are big and sluggish and so supply a lot less outward force than the fast-moving electrons. But why are there free electrons about anyway? Normally, in a cold gas – and remember, the star no longer has any internal fires – all the free electrons are tucked up in bed around their nuclei. The answer is they are so close together that the orbits of electrons are bigger than the separation of the nuclei. In the jargon, they are 'pressure ionised'.

13   Although there have now been three Nobel Prizes for Physics awarded for work on pulsars, none has gone to their discoverer, Jocelyn Bell.

14   Terry Pratchett, *Small Gods*, Corgi, London, 2013.

15   Dan Simmons, *The Fall of Hyperion*, Gollancz, London, 2005.

16   The term 'big bang' was coined by the British astrophysicist Fred

Hoyle during a BBC radio broadcast in 1949. Ironically, Hoyle, one of the creators of the alternative 'Steady State' theory in 1948, never believed in the big bang.

17 Cepheid variables have the property, discovered by Henrietta Leavitt in 1908, that the greater their pulse period the greater their intrinsic luminosity. This means it is always possible to deduce from a Cepheid's period its true luminosity. Knowing how bright it appears from Earth, astronomers can then ask: how far away would it have to be to appear as faint as it does?

18 'Space is big. Really big. You just won't believe how vastly, hugely, mind-bogglingly big it is.' Douglas Adams, *The Hitchhiker's Guide to the Galaxy*, Chapter 8.

19 In exactly the same way that the frequency, or 'pitch', of a police siren gets higher as it approaches and lower as it recedes, the frequency of light emitted by atoms in a star becomes higher or lower depending on whether the star is approaching or receding from the Earth. By measuring the magnitude of this 'Doppler shift' in frequency for common atoms of elements such as hydrogen, astronomers can determine the velocity of a star's movement towards us and away from us.

20 The reason that the building of elements requires a high temperature is that the nuclei of atoms are positively charged. Like charges repel, so two nuclei have a ferocious resistance to being pushed together. But, if they are slammed together at high enough speed, this repulsion can be overcome and they can approach close enough to be grabbed by the 'nuclear force' and stick together. High speed is synonymous with high temperature, temperature being nothing more than microscopic motion.

21 Arthur Eddington, who had invented the theory of stellar interiors, believed that, as stars rotate, currents of gas circulating in their interiors thoroughly mix their gas. This means that as a star turns its hydrogen fuel into helium ash, the by-product of which is sunlight, the hydrogen is spread throughout the star, becoming ever more dilute until the star's fires dim and go out. Eddington was wrong. There is no such mixing. Instead, helium ash falls to the centre of the star, where it is compressed and heated. When hydrogen in the core is exhausted, helium may burn to carbon, which in turn falls to the centre of the star and is compressed and heated. The upshot is that stars, far from being uniformly mixed and going out with a whimper, become 'chemically differentiated', their interiors supporting ever more extreme temperatures and densities for millions or billions of years. Exactly the kind of furnace needed to build up elements.

22 The main reason the big bang fireball could not forge all the elements is that it expanded and cooled too quickly. The conditions necessary

for element building existed from about a minute after the moment of creation until the Universe was about 10 minutes old. Nevertheless, there was time to forge the lightest elements like lithium and beryllium. In particular, the fireball of the big bang could convert 10 per cent of the hydrogen nuclei into helium nuclei. This is exactly the proportion astronomers observe in the Universe today and is taken as strong evidence for the big bang. Pretty much all the heavier elements – from the iron in your blood and the calcium in your bones to the oxygen you take in with every breath – have been forged in stars since the big bang.

23 A black body absorbs all the heat that falls on it. The heat is distributed between all the atoms by countless collisions in which fast-moving atoms transfer energy to slower-moving atoms. The result is that the black body emits heat that depends in no way on the kind of atoms the body is made of. Instead, 'black body radiation' has a universal spectrum which depends only on one number: the body's temperature.

24 Marcus Chown, *Afterglow of Creation*, Faber & Faber, London, 2010.

25 L. S. Schulman, 'Source of the observed thermodynamic arrow', *Journal of Physics: Conference Series*, vol. 174, 2008, p. 12,022.

26 Einstein himself never believed in black holes. In fact, in October 1939 he published a paper in which he claimed (incorrectly) that, for a black hole to form from a collection of stars, they would have to orbit each other faster than the speed of light, which is forbidden by the special theory of relativity. See Albert Einstein, 'On a Stationary System With Spherical Symmetry Consisting of Many Gravitating Masses', *Annals of Mathematics*, Second Series, vol. 40, No. 4, 1939, p. 922 (http://www.jstor.org/stable/1968902).

27 See Marcus Chown, *Quantum Theory Cannot Hurt You*, Faber & Faber, London, 2014.

28 Strictly speaking, quantum theory is not a theory of small things but a theory of 'isolated' things – that is, things not influenced by their surroundings. In practice, though, this makes quantum theory a theory of small things because it is easy to isolate an atom from its surroundings but hard to isolate, for instance, a human being like you. Molecules of air and particles of light are continually bouncing off you.

29 Albert Einstein, 'Naherungsweise Integration der Feldgleichungen der Gravitation' [Approximate integration of the field equations of gravitation], *Sitzungsber der Preussische Akademie der Wissenschaften*, 22 June 1916, p. 688. (Also in *The Collected Papers of Albert Einstein, vol. 6, The Berlin Years: Writings, 1914–1917*, Princeton University Press, 1997, p. 201.)

Chapter 8

1   William Bragg, 'Electrons and Ether Waves (The Robert Boyle Lecture 1921)', *Scientific Monthly*, vol. 14, 1922, p. 158.

2   Said by Niels Bohr to Wolfgang Pauli after his presentation of Heisenberg's and Pauli's nonlinear field theory of elementary particles, Columbia University (1958), as reported in Freeman Dyson, 'Innovation in Physics', *Scientific American*, vol. 199, No. 3, September 1958, p. 74.

3   'Gamma rays' are even more energetic than X-rays. They were discovered by the French chemist and physicist Paul Villard in 1900 and named by New Zealand physicist Ernest Rutherford in 1903. Gamma rays come from inside the atomic nucleus, which is the seat of enormous energies.

4   When light is shone on certain metals, electrons are ejected from their surface. Increasing the amount, or 'intensity', of the light liberates more electrons. But, if the light has less than a threshold energy, no electrons are emitted, no matter how intense the light. As Einstein realised, this 'photoelectric effect' is explained if light consists of photons, and only photons of sufficient energy can kick out electrons from the metal.

5   In fact, once the existence of atoms was confirmed, and they were shown to be so small that 10 million would span the full stop at the end of this sentence, there was the paradox that the wavelength of visible light is about 10,000 times bigger than an atom. There seems no way an atom can absorb or spit out light of that size – unless light, like an atom, is a small and localised thing – a photon.

6   According to the standard picture of cosmology, known as 'inflation', the Universe started out so ultra-tiny that it contained hardly any information. Today, by contrast, it contains a truly vast amount – just imagine how much is needed to describe the type and location of every atom in the Universe. The puzzle of where all the information came from is explained by quantum theory since randomness is synonymous with information. Every random quantum event since the big bang, such as the decay of a radioactive atom, has injected information/ complexity into the Universe. When Einstein said, 'God does not play dice with the Universe', he could not have been more wrong. If God had not played dice, there would be no Universe – certainly no Universe with anything interesting going on in it. See the chapter 'Random Reality' in Marcus Chown, *The Never-Ending Days of Being Dead*, Faber & Faber, London, 2007.

7   See Chapter 7.

8   Werner Heisenberg, *Physics and Philosophy*, Penguin Classics, London, 2000.

9   Interference is a defining feature of waves. If two waves overlap,

where the peaks of the two waves coincide, they boost each other, or 'constructively interfere', and where the peaks of one wave coincide with the troughs of the other wave, they cancel each other out, or 'destructively interfere'. It is precisely this effect which was demonstrated for light by Thomas Young in 1801. (See Chapter 5.)

10 Strictly speaking, the probability of finding a particle at any location is the square of the amplitude wave at a particular location. The probability is always a number between 0 and 1, with 0 corresponding to 0 per cent probability and 1 corresponding to 100 per cent probability.

11 Most physicists believe that quantum systems are isolated systems and that they stop behaving in a quantum manner because of a process called 'decoherence'. The key thing to understand is that we never actually see quantum behaviour directly. When a photon is detected by the human eye, for instance, the photon leaves an impression on hundreds of atoms. It is this impression that the brain observes (so, in a sense, we only ever observe ourselves!). And it is because it is very hard to keep hundreds of atoms in a superposition – the waves stop overlapping, or decohere – that quantumness is lost. The flip side of this is that, if it were possible to keep all those atoms in a superposition, quantumness could in principle manifest itself at any size. Currently, physicists are trying to do that. In a 'quantum computer', they want to exploit the ability of quantum systems to do many things at once, to do many calculations at once. Roger Penrose, however, believes that quantumness cannot manifest itself at any size and there is a threshold mass beyond which there is a transition from quantum to classical physics. The question of who is right may have to be resolved by experiment. See Marcus Chown, *Quantum Theory Cannot Hurt You*, Faber & Faber, London, 2006.

12 In fact, the problem of reconciling the quantum world, where things exist as a haze of probabilities, and the everyday world, where things exist with certainty, is fundamental and mysterious. There are at least thirteen 'interpretations' of quantum theory that attempt to do this, all of which predict the same outcome for every known experiment. Perhaps the most mind-blowing is the one proposed by Hugh Everett III in 1957. According to the 'Many Worlds' interpretation, the separate waves in a superposition actually describe separate arms of reality. So, in the case of the oxygen atom in superposition of two waves, one describing it on the left-hand side of a room and the other on the right-hand side, the oxygen atom really is in two places at once, one in one parallel reality and the other in a separate parallel reality.

13 This is a remarkable discovery of the French mathematician Joseph Fourier (1767–1830), who found that, by superposing sine waves of different wavelengths and different 'phases' (that is, different locations

of their peaks relative to each other), it was possible to create a wave
of any shape whatsoever – even a square, top-hat-shaped one. One way
of putting this is that, just as atoms are the basic building blocks of all
matter, sine waves are the basic buildng blocks of all waves.

14   See Chapter 8.

15   Heisenberg himself came up with another explanation of the
      Uncertainty Principle, claiming that the wave nature of anything used
      to 'see' an object made it impossible to know exactly where it was.
      This is what countless university physics students are taught. But
      Heisenberg was incorrect. The Uncertainty Principle has nothing to do
      with measurement. The uncertainty is intrinsic to the submicroscopic
      world. See Geoff Brumfiel, 'Quantum uncertainty not all in the
      measurement: A common interpretation of Heisenberg's uncertainty
      principle is proven false', *Nature*, 11 September 2012.

16   Consider a wave packet made of light that passes by. Because there
      is an uncertainty in its location (dx), it follows that there is an
      uncertainty in the exact time it passes (dt) equal to dx/c, where c is
      the speed of light. And because there is an uncertainty in momentum
      (dp), it follows that there is an uncertainty in its energy (dE) equal
      to $dp \times c$. Since $dp \times dx > h/2\pi$, it follows that $dE \times dt > h/2\pi$. In this
      case, the wave is (conveniently) travelling at the speed of light. But the
      result can also be shown to be true for a more general wave packet
      representing a quantum particle – though a demonstration would be
      more complicated.

17   The quantum vacuum is an unavoidable consequence of two things,
      the first of which is the existence of fields of force. As pointed out,
      physicists view fundamental reality as a vast sea of such fields. In their
      picture, known as 'quantum field theory', the fundamental particles
      are mere localised humps, or knots, in the underlying fields. The best
      understood of all the fields, and the one with the greatest bearing on
      the everyday world because it glues together the atoms in our bodies
      – not to mention all other normal matter – is the electromagnetic
      field. The electromagnetic field can undulate in an infinite number of
      different ways, each oscillation 'mode' corresponding to a wave with a
      different wavelength. Think of the waves at sea, which can range all the
      way from huge, rolling waves down to tiny ripples. Naively, the vacuum
      of empty space would be expected to contain no electromagnetic waves
      whatsoever. And this would be true but for the small matter of the
      Heisenberg Uncertainty Principle. According to the principle, every
      conceivable oscillation of the electromagnetic field must contain at
      least a minimum amount of energy. This seemingly innocuous rule
      has dramatic and profound implications for the vacuum because it
      means that each of the infinite number of possible oscillation modes
      of the electromagnetic field must be jittering with the minimum energy

dictated by the uncertainty principle. In other words, the existence of each mode is not simply a possibility, it is a certainty. Far from being empty, the 'quantum vacuum' has an extraordinary energy density – far greater than even inside an atomic nucleus. The reason we do not notice it is the same reason we do not notice the air: *because it is the same everywhere.*

18   Vlasios Vasileiou et al., 'A Planck-scale limit on space-time fuzziness and stochastic Lorentz invariance violation', *Nature Physics*, vol. 11, 2015, p. 344 (http://www.nature.com/nphys/journal/v11/n4/full/nphys3270.html); Eric Perlman et al., 'New constraints on quantum gravity from X-ray and gamma-ray observations', *Astrophysical Journal*, vol. 805, No. 1, 20 May 2015, p. 10 (http://arxiv.org/pdf/1411.7262v5.pdf).

19   Natalie Wolchover, 'Visions of Future Physics', *Quanta Magazine*, 22 September 2015 (https://www.quantamagazine.org/20150922-nima-arkani-hamed-collider-physics/).

20   Max Planck, 'Über irreversible Strahlungsvorgänge', *Annalen der Physik*, vol. 4(1), 1900, p. 122.

21   Tony Rothman and Stephen Boughn, 'Can gravitons be detected?', 2008 (http://arxiv.org/pdf/gr-qc/0601043.pdf).

22   An electronvolt (eV) is the energy gained by an electron after being accelerated by 1 volt. A gigaelectronvolt (GeV) is an energy 1 billion times bigger.

23   Tushna Commissariat, 'BICEP2 gravitational wave result bites the dust thanks to new Planck data', *Physics World*, 22 September 2014 (http://physicsworld.com/cws/article/news/2014/sep/22/bicep2-gravitational-wave-result-bites-the-dust-thanks-to-new-planck-data)

Chapter 9

1   *Preussische Akademien der Wissenschaften, Sitzungsberichte*, Berlin, 1916, p. 688.

2   Douglas Adams, *The Restaurant at the End of the Universe*, Pan Books, 1980.

3   'Why is quantum gravity so hard? And why did Stalin execute the man who pioneered the subject?', *Scientific American* guest blog, 14 July 2011 (http://blogs.scientificamerican.com/guest-blog/why-is-quantum-gravity-so-hard-and-why-did-stalin-execute-the-man-who-pioneered-the-subject/).

4   Matvei Bronstein, 'Vsemirnoe tyagotenie i elektrichestvo (novaya teoriya Eynshteyna)' ['Universal gravity and electricity (new Einstein theory)'], *Chelovek i priroda*, vol. 8, 1929, p. 20.

5   A particle with spin 2 looks the same if you rotate it through half a turn. Think of a double-headed arrow. A particle with spin 1 looks the same after if you rotate it through 1 turn. Think simply of a normal

arrow. But a particle with spin ½ looks the same only after it has been rotated through two turns! Say you were not the same person if you turned round once but only if you turned round *twice*. Well, that is the way it is for electrons, the most common example of a particle with spin ½. If quantum spin is something new under the sun, spin ½ is something doubly new under the sun.

6   See 'No more than two peas in a pod at a time', Marcus Chown, *We Need to Talk About Kelvin*, Faber & Faber, London, 2009.

7   Special relativity and quantum theory also impose a tight constraint on how particles interact via a force-carrier. You might imagine that a particle can interact simultaneously with five or twelve or any number of force-carriers. But, actually, it can interact with only one. The space-time diagram commonly used to depict such an event is known as a Feynman diagram. And, on a Feynman diagram, the restriction is equivalent to only three particles meeting at a space-time point, or 'vertex'. For instance, an electron comes into a vertex, a photon meets it and is absorbed, and an electron (redirected, or 'scattered') flies outwards. But special relativity and quantum theory simplify things only in the normal, low-energy/long-range world. In the high-energy/short-range world of quantum gravity there is sufficient energy for the more complex interactions.

8   Steven Weinberg, *The Quantum Theory of Fields*, Cambridge University Press, Cambridge, 2005.

9   See Chapter 8.

10   The best candidate for the dark matter is the lowest-mass supersymmetric particle. The 'neutralino' is in fact a superposition of three particles – a photino, a Higgsino and a Z-ino.

11   One of the biggest unsolved mysteries is why we live in a matter-dominated Universe. The best guess of physicists is that, in the big bang, some lop-sidedness in the laws of physics either favoured the creation of matter or preferentially destroyed antimatter.

12   Gottfried Leibniz, *Discours de métaphysique*, 1686.

13   The Polish-born Nobel prize-winner actually said, 'Who ordered that?' on the discovery of the muon, a heavy version of the electron, in 1936.

14   A rival, but more conservative, approach to finding a deeper theory than Einstein's theory of gravity is called 'loop quantum gravity': see Lee Smolin, *Three Roads to Quantum Gravity*, Basic Books, London, 2002. The theory describes gravity at the quantum scale but makes no attempt to unify it with the other forces. Also, no one has yet been able to show that it leads to the general theory of relativity on the large scale.

15   Actually, quarks are confined in two distinct configurations. A triplet of quarks makes a 'baryon', the most common of which are the proton and neutron; whereas a quark–antiquark pair makes a 'meson'. Quarks

are actually confined within baryons and mesons only at low energies. At ultra-high energies such as those that existed in the first moments of the big bang, they can break free of their prisons to form an amorphous 'quark-gluon' plasma.

16   Since gravity leaks out in all directions, at a distance r from a mass, its effect is spread out over the surface of a sphere of area $4\pi r^2$ and so diluted by $1/4\pi r^2$. This is the origin of gravity's inverse-square-law force.

17   This is exactly what happens to a magnetic field inside a 'superconductor', a material cooled to a temperature at which its electrical resistance vanishes. Within the material, the magnetic field is confined to narrow channels known as 'flux tubes'.

18   Gabriele Veneziano, 'Construction of a crossing-symmetric, Regge-behaved amplitude for linearly rising trajectories', *Nuovo Cimento A*, vol. 57, 1968, p. 190. Veneziano's theory was called the 'dual resonance model' and only later became known as string theory.

19   Roy H. Williams, 'String Theology', 31 July 2006 (http://www. mondaymorningmemo.com/newsletters/string-theology/).

20   'The mind-blowing concepts of one of the world's most brilliant theoretical physicists', Australian Broadcasting Corporation, 25 February 2016.

21   Arthur C. Clarke, 'The Wall of Darkness', *The Other Side of the Sky*, Gollancz, London, 2003.

22   Lisa Randall and Raman Sundrum, 'Large mass hierarchy from a small extra dimension', *Physical Review Letters*, vol. 83 (17), 1999, p. 3,370 (http://arxiv.org/pdf/hep-ph/9905221v1.pdf); Lisa Randall, *Warped Passages: Unravelling the Mysteries of the Universe's Hidden Dimensions*, HarperCollins, New York, 2006.

23   The radius of the horizon of black holes goes up in step with their mass. So a black hole with twice the mass of another has a horizon of twice the radius. But, because the force of gravity weakens according to the square of distance, this means a black hole with twice the mass of another actually has a gravity only half as strong. Not only this but the rate at which the gravity of such a hole changes – the 'tidal force' – is only one-quarter as strong. Since it is the tidal force that ultimately tears apart particle–antiparticle pairs, creating Hawking radiation, Hawking radiation is comparatively weak for big black holes and strong for small black holes.

24   Steve Connor, 'Stephen Hawking admits the biggest blunder of his scientific career – early belief that everything swallowed up by a black hole must be lost for ever', *Independent*, 11 April 2013 (http://www. independent.co.uk/news/science/stephen-hawking-admits-the-biggest-blunder-of-his-scientific-career-early-belief-that-everything-8568418. html).

25   A black body absorbs all the heat that falls on it. The heat is distributed between all the atoms by countless collisions in which fast-moving atoms transfer energy to slower-moving atoms. The result is that the black body emits heat that depends in no way on the kind of atoms the body is made of. Instead, 'black body radiation' has a universal spectrum that depends only on one number: the body's temperature.

26   Jacob Bekenstein, 'Black holes and the second law', *Nuovo Cimento Letters*, vol. 4, 1972, p. 737; Jacob Bekenstein, 'Black holes and entropy', *Physical Review D*, vol. 7, 1973, p. 2,333.

27   Andrew Strominger and Cumrun Vafa, 'Microscopic origin of the Bekenstein–Hawking entropy', 1996 (http://arxiv.org/pdf/hepth/9601029v2.pdf).

28   Although the Universe is 13.82 billion years old, the distance to the cosmic light horizon – the edge of the observable Universe – is about 42 billion light years. This is because the Universe, during its first split-second of existence, 'inflated' far faster than the speed of light. This does not violate relativity because space – the backdrop to cosmic events – can expand at any rate whatsoever.

29   Juan Maldacena, 'The Large N Limit of Superconformal field theories and supergravity', *Advances in Theoretical and Mathematical Physics*, vol. 2, 1998, p. 231 (http://arxiv.org/pdf/hep-th/9711200.pdf).

30   See Chapter 8.

31   Van Raamsdonk, quoted in Ron Cowen, 'The quantum source of space-time', *Nature*, vol. 527, 19 November 2015, p. 290.

32   Albert Einstein, Boris Podolsky and Nathan Rosen, 'Can quantum-mechanical description of physical reality be considered complete?', *Physical Review*, vol. 47 (10), May 1935, p. 777 (http://journals.aps.org/pr/pdf/10.1103/PhysRev.47.777).

33   Albert Einstein and Nathan Rosen, 'The particle problem in the general theory of relativity', *Physical Review*, vol. 48 (1), July 1935, p. 73.

34   Light is given out when an electron in an atom drops from a high-energy to a low-energy orbit. No light is given out by atoms such as hydrogen – which each possess a single electron – if either the atoms are so cold that every electron is in its lowest-energy orbit or the atoms are so hot that they have been stripped of their lone electrons.

35   Repulsive gravity can come about because in the general theory of relativity the 'source' of gravity is in fact energy density (u) + 3 × pressure (P). The pressure exerted by the atoms of normal matter is negligible compared with the energy density of matter. But there is the possibility of novel 'stuff' where this is not true. The dark energy is such stuff. In fact, for the dark energy, the pressure is not only negative – it sucks rather than blows – but less than $-1/3u$. This

reverses the 'sign' of the source of gravity, turning it from attractive to repulsive. It is this repulsive gravity that is speeding up the expansion of the Universe. The irony is that the dark energy is everywhere trying to shrink. Only through general relativity does this manifest itself as repulsive gravity.

36   In the general theory of relativity, empty space can have intrinsic curvature, or energy. This is known as the 'cosmological constant'. Zero is a very special number and cosmologists have not been surprised to find that the cosmological constant is non-zero. The surprise is its smallness. Quantum theory predicts that, because of 'quantum fluctuations', the vacuum should contain energy. But quantum theory predicts an energy-density for the vacuum – that is, a value for the dark energy – which is 10^{120} (1 followed by 120 zeroes) bigger than what is observed. This is the biggest discrepancy between a prediction and an observation in the history of science! This number could be brought down to the energy density actually observed if there is another contribution to the vacuum energy which is negative and differs only in the 119th decimal place. This is a tall order. But it is conceivable that supersymmetry could do this since the energy of the fluctuations in the boson fields is positive while the energy in the fermion fields is negative.

37   Mordehai Milgrom of the Weizmann Institute in Rehovot, Israel, believes that, below an acceleration of about one-billionth of a g, gravity changes to a stronger form that does not weaken as quickly with distance as an inverse-square-law force. This Modified Newtonian Dynamics, or MOND, can describe the motions of stars orbiting in all spiral galaxies with a single formula. By comparison, a different amount of dark matter with a different distribution is required to explain the motion of stars in each spiral galaxy. A form of MOND which is compatible with Einstein's theory of relativity was found by Jacob Bekenstein of the Hebrew University of Jerusalem in 2000. See Jacob Bekenstein, 'Relativistic gravitation theory for the MOND paradigm' (http://arxiv.org/pdf/astro-ph/0403694v6.pdf).

38   Rory Carroll, 'Kip Thorne: physicist studying time travel tapped for Hollywood film', *Guardian*, 21 June 2013 (https://www.theguardian.com/science/2013/jun/21/kip-thorne-time-travel-scientist-film).

39   B. Oberg (ed.), *The Papers of Benjamin Franklin*, vol. 31, Yale University Press, New Haven, 1995, p. 455.

40   'Clarke's Third Law', *Profiles of the Future*, Gateway, London, 2013.

Index